BFI TELEVISION MONOGRAPH

8

Television
and History

Colin McArthur

produced by
The British Film Institute
Educational Advisory Service

General Editor:
Edward Buscombe

British Film Institute,
127 Charing Cross Road, London WC2H 0EA
1978

The past, whether presented through actuality footage or dramatisation, is a recurrent source of subject matter for television programme makers. Despite the apparent variety of such history programmes, the author of this monograph argues, they are shaped primarily within the categories of *television* rather than the needs of *historical knowledge.* And to the extent that they do have a conception of history, it is one which is outdated and discredited but which meshes very easily with the categories of television to facilitate a particular construction of the social world, a construction which is uncongenial to radical change.

Following a critique of several such history programmes, the author points towards possible alternative models.

Dedication

To all television workers — makers, critics and teachers — seeking alternative forms . . .

The Author

Colin McArthur was born in Glasgow in 1934. Having left school at fifteen he served a craft apprenticeship, did National Service in the Army and had a range of casual jobs before entering university. After graduation he taught in secondary, further and higher education before joining the BFI's Education Department in 1968. He was Teacher Adviser, then Editor of Film Study Materials before becoming Head of the BFI's Film Availability Services in 1974.

He has contributed to *Screen, Screen Education, Film Form* and *Sight and Sound,* is the author of *Underworld USA* (a study of the American crime film) and wrote a chapter in *Football on Television,* Monograph No. 4 in this series. Since 1972 he has written on film, television and general arts policy for *Tribune.*

Contents

Foreword

If there are skills deployed in the construction of this monograph they are assuredly not those of the professional historian. Consequently, anyone seeking in it a discussion of historical 'accuracy' or 'interpretation' in this or that programme is going to be disappointed.

What the monograph does is to advance a very broad, polemical and, within the Marxist perspective, non-sectarian argument about possible relationships among the cluster historiography/television/ideology. The main focus of interest is in the *scope* rather than the *nuance* of the argument, although an attempt is made to demonstrate relationships empirically. This has meant that something less than justice has been done to the complexity of many of the issues discussed herein: for example the immense diversity and (sometimes byzantine) richness of the debate among Marxists on the relationship between base and superstructure and the nature of the relationship between the State and television institutions. However, it is necessary sometimes to risk vulgarisation and eclecticism for the sake of generality and accessibility.

The monograph is written less *by* the person whose name appears on the cover than *through* him in the sense that it would have been inconceivable without an ongoing process of debate involving many groups and individuals by no means all of whom subscribe to the theoretical and political positions evident herein. Some of the stations of this debate are the British Film Institute's Film Availability Services and Educational Advisory Service; the journals and weekend schools of the Society for Education in Film and Television; the annual seminar at the Edinburgh Film Festival; and, beyond these, the discussions which have been informing British Marxism over the last decade.

To revert to a more traditionally authorial note, many thanks are due to Paul Madden, Jeremy Boulton, Clive Truman and Tim Cotter for arranging and servicing the viewing of much of the material discussed herein; to Joy Wong for typing the manuscript; to Ed Buscombe and Angela Martin for pointing out some of its grosser impenetrabilities; and to Erich Sargeant for preparing the photographs from which the photomontages were constructed.

1 The Concept of Ideology

Don't you think there could be greater security in the love of your people than in the strength of your police force? (Edward VII, in the ATV series of that name, to Czar Alexander III)

The intervention this monograph seeks to make hinges on the concept of *ideology*,[1] yet few concepts are more generally misunderstood at this particular historical moment due to the different meanings accorded the term 'ideology' in different discourses.

In popular usage (and in the definitions in many dictionaries and encyclopaedias of social science) the term has heavily pejorative connotations and describes sets of ideas which are explicit, overtly propagandistic and — above all — directly concerned with politics: e.g., 'the ideology of communism', or 'the ideology of fascism'. Implicit in such a usage is the suggestion that the available twentieth century option to 'ideologies' thus conceived, liberal democracy, is a *non*-ideological (and natural) condition — a notion that is *highly* ideological in the other sense of the term.

This other usage — which incorporates and goes beyond the above meaning — is that of Marxist discourse. Ideology therein conceived, far from describing a condition in which the ideas which underpin and give the appearance of coherence to social experience are explicit, describes the very reverse condition. As V. L. Allen puts it:

> . . . ideology is the process through which ideas, values and purposes act to influence behaviour . . . therefore, it is present in all societies at all times. The ideas, values and purposes have to be articulated in a particular way in order to influence behaviour. They have to be presented so as to enter into the thinking of people; to appear through analytical tools which people handle in their everyday lives without really knowing they are doing so; they have to make their impact through the explanations which ordinary people reach about their own and others' situations. They have therefore to be embodied in theories. Theories form an integral part of ideologies. Put differently, an ideology is in effect about the dissemination of theories.[2]

Ideology thus conceived, far from being restricted to explicit and programmatic statements by 'extremist' political groups (as the dominant discourse would have it) is a feature of every society at every historical moment, the 'social cement', in Gramsci's terms, whereby the power of dominant groups is maintained without regular and widespread recourse to physical coercion.

Marxism, analysing successive kinds of society (slave, feudal and capitalist) describes the social formation as rather like the several floors of a house, the ground floor (the *base* or *infrastructure*) consisting of the forces of production and the social relations of production and the upper floors (the *superstructure*) consisting first of all of the coercive factors of the state and the legal system and, beyond that, of the factors concerned primarily with lived experience — the family, religion, education, the arts and the mass media.

The simile of the successive levels of a house is a particularly pertinent one since it poses the notion of a *structural* relationship among the various levels with the 'ground floor' being the most important in that it holds the others up.

All Marxians accept some form of this analytic model although there is considerable debate about the nature and extent of the dependence of the *superstructure(s)* upon the *base*: some argue that the economic base is very directly *reflected* in the superstructure(s), others that the superstructure(s), though 'in the last analysis' dependent on the base, have a great degree of autonomy. This autonomy has the consequence of masking the relationship with the base and of allowing it to be reciprocal, dialectical, rather than the base simply acting in a one-way direction on the superstructure(s).

The most interesting work of the latter tendency is that of Antonio Gramsci, Louis Althusser and Nicos Poulantzas[3] and the present monograph is an attempted intervention in the same tradition. In particular, it leans very heavily on the work of Louis Althusser though it doubtless crudifies that work in the interests of presenting a broad and accessible argument concerning the relationships among the cluster historiography/television/bourgeois society*. While all of Althusser's work is relevant to the question of ideology, the key essay for the purposes of this monograph is 'Ideology and the Ideological State Apparatuses (Notes towards an Investigation)'.[4]

Being concerned with a general theory of the superstructure, Althusser refines the first level of the superstructure from the classical Marxist 'state apparatus' of government and legal system to the 'Repressive State Apparatus' since it functions primarily by coercion — administrative and, if necessary, physical. The second level of the superstructure Althusser characterises as the site of the 'Ideological State Apparatuses' (hereafter — following Althusser — referred to as ISAs) which function primarily by the 'packaging' of consciousness (the lived experience mentioned above) in ways congenial to the maintenance of the existing relations of production in the economic base.

Althusser instances several such ISAs: the religious ISA; the educational ISA; the family ISA; the legal ISA; the political ISA; the trade union ISA; the communications ISA; and the cultural ISA. It is important to remember (as the earlier quotation from V. L. Allen indicates) that the concept of *ideology* does not imply any kind of conscious, cynical manipulation within the ISAs in maintaining the existing relations of production. On the contrary, the real function of ideology is masked from itself and its agents. It is of the essence

---

*'Bourgeois society' is used here as a descriptive term to refer to the form of society — characterised by the rise to power of primarily urban groups associated with mercantile, finance and industrial capital and state bureaucracy — which superseded feudal society.

of ideology that within it historically determined ideas and situations are rendered *natural, obvious* and *axiomatic,* become the small change of *common sense,* honestly subscribed to by many people.

For example, throughout the continuing economic crisis in Britain the phrase 'X must get a reasonable return for his/her/its investment' is repeated. No 'reasonable' person is heard to dissent from this sentiment: it seems the most 'obvious' and 'natural' thing to say, a 'fact' of 'human nature'. Yet approval of the practice of *usury,* among 'Christians' at any rate, is relatively recent, dating quite precisely from the early to mid-sixteenth century, with certain papal decrees and scriptural commentaries by John Calvin signalling the shift in attitude.[5] It is difficult now to conceive that the above sentiment — so 'natural' to our society — would have evoked in medieval Christendom a response perhaps analogous to our response to child molestation.

The economic ideology of 'X must get a reasonable return for his/her/its investment' — and its demotic variant 'a fair day's work for a fair day's wage' — is clearly congenial to the existing system of production relations in capitalist society. It could be demonstrated that the *dominant* ('dominant' because the contradictions in the society permit a limited currency of alternative ideologies) ideologies of the ISAs outlined above are equally congenial to the maintenance of that system. For example, the separation of the social and the religious inherent in the phrase 'render unto God what is God's and unto Caesar what is Caesar's' has traditionally ensured that awkward contradictions between the teachings of Christ and the practice of Christianity in capitalist societies have remained masked or have at least become manageable. The construction of children as 'individuals' in competition with each other within the educational ISA is highly congenial to the continuation of the existing system of production relations and, indeed, provides a major mechanism for servicing the various levels of that system. So too the ideology of 'balance and impartiality' (which incidentally, empirical evidence is revealing as less and less tenable) in the television sector of the communications ISA is supportive of the existing system of production relations by being 'balanced' and 'impartial' with regard only to those who operate within the consensus view that the capitalist system is inherently sound and, subject to periodic 'reforms', ought to continue indefinitely.[6]

This indicates the *systemic* nature of bourgeois ideology. As John Mepham puts it:

> Ideology is not a collection of discrete falsehoods but a matrix of thought firmly grounded in the forms of our social life and organised within a set of interdependent categories. We are not aware of these systematically generative interconnections because our awareness is organised through them.[7]

The process pointed to by Mepham has been investigated in the work of the French psychoanalyst Jacques Lacan, work which, in its turn, has been highly influential in discourse about film and television in Britain. The nature of Lacan's project and its relevance for work on the media has been well described by John Caughie:

> It has been the further value of the intervention of Jacques Lacan and the insertion of psychoanalysis into the problematic that ideology has come to be construed not as a product of 'false consciousness' but as a process of the construction of human subjects, and of assigning these subjects to positions within a (social) formation. The importance of representation within such a view is apparent, and the claim that representations are determinant can be seen not as the simplistic notion that people believe what they are told, but as a more elaborated model in which people are placed in positions in a particular kind of world — the 'real world' — through its representations.[8]

This quotation indicates the political urgency of proceeding with work on the relationship between ideology and representation over the whole range of television, a project which provides the context of this monograph.

It is wholly in accord with the notion of an *attenuated* relationship between *base* and *superstructure(s)* advanced in this monograph that the perception of reciprocal connections between the relations of production and the ISAs should be difficult. The connection between individual men and women buying or selling labour power through the market mechanism and, say, the forms of television programmes dealing with history both as 'fact' and 'fiction' is not immediately apparent if sought as some kind of direct one-to-one relationship. (Nor, as will be argued presently, is the relationship between the method of bourgeois historiography and modes of television production.) However, the connection might become more apparent if the problem is posed as the search for a set of common *philosophical* assumptions underpinning both activities.

The emergence of the capitalist mode of production with its characteristic economic relations whereby individual men and women are 'free' to sell their own labour power or buy that of others required a concurrent philosophical definition of men and women as separate and autonomous entities, precisely as *individuals.* Just as in our society the notion of usury has become naturalised, regarded as a timeless 'fact of life', so too has the notion of men and women as individuals. In fact, this notion had to be laboriously constructed as an ideology of the rising bourgeoisie to render coherent its throwing off of feudalism. Nicos Poulantzas, making the different point that in any social formation a particular region of ideology becomes dominant, nevertheless writes:

> The dominant form under which the bourgeois class experienced its first protests against the feudal order and experienced its subsequent conditions of existence, and which has permeated the ensemble of capitalist formations is the form of juridico-political discourse. Liberty, equality, rights, duties, the rule of the law, the legal state, the nation, individuals/ persons, the general will, in short all the catchwords under which bourgeois class exploitation entered and ruled in history were directly borrowed from the juridico-political sense of these notions, as formulated

4

for the first time by medieval legal theorists of the social contract in the Italian universities.[9]

At the directly philosophical level, this process was reinforced by the constitution of the knowing subject in the work of the seventeenth century French philosopher Descartes and in the primacy given to sense-data as perceived by the individual consciousness in the work of the seventeenth century English philosopher Locke. It might also be argued that the constitution of, individual perception was recognised by the emergence of perspective in Renaissance Italian painting, by the freezing of art from the point of view of the discrete individual allotted his/her space as such and contemplating the world as though standing at an open window:

> It was a question for a society in process of total transformation of a space in accordance with its actions and its dreams . . . It is men who create the space in which they move and express themselves. Spaces are born and die like societies; they live and have a history. In the fifteenth century, the human societies of Western Europe organised, in the material and intellectual sense of the term, a space completely different from that of the preceding generations; with their technical superiority they progressively imposed that space over the planet.[10]

Thus in late medieval and Renaissance Europe, the discourses of Law and Politics, of Philosophy and Art were buckling and reforming round one pre-eminent category.

It will be argued in this monograph that precisely this ideological category of *the individual*, which provides the philosophical basis for the relations of production under capitalism, is one of the key categories for understanding both the dominant forms of history-writing in Britain and the forms of television programmes about history.

The function of this, as of every, ideological category is to offer an explanation — highly plausible because of its closeness to the contours of the phenomenal world as perceived by the senses — of the nature of history and contemporary social experience and to forbid entry to the *real* categories (e.g., class, mode of production) which shape our lives.

Notes

1. The concept of ideology is discussed here in terms of its contemporary usage. The historical origins and development of the term and the place of the *concept* of ideology in the major post-French Revolution philosophical positions are usefully outlined in George Lichtheim's 'The Concept of Ideology', in *History and Theory*, IV, 2 (1965).
2. V. L. Allen, *Social Analysis: a Marxist Critique and Alternative*, Longman, London 1975, p.24.
3. Antonio Gramsci, *The Prison Notebooks*, Lawrence and Wishart, London 1971;

Louis Althusser, *For Marx*, Allen Lane, The Penguin Press, London 1969, and *Reading Capital*, New Left Books, London 1970; Nicos Poulantzas, *Political Power and Social Classes*, New Left Books and Sheed & Ward, London 1973.

4. In *Lenin and Philosophy*, New Left Books, London 1971, pp.123-173.

5. See Benjamin Nelson, *The Idea of Usury*, University of Chicago Press, Chicago 1969 (2nd. edn.).

6. For an excellent analysis of precisely this issue of balance and impartiality and an attempt to indicate concretely what the notion of 'relative autonomy' might mean, see Stuart Hall, Ian Connell and Lidia Curti, 'The "Unity" of Current Affairs Television' in *Working Papers in Cultural Studies*, No. 9, Centre for Contemporary Cultural Studies, University of Birmingham, Spring 1976.

7. 'The Theory of Ideology in *Capital*', in *Radical Philosophy* No. 2, Summer 1972. Mepham's article also gives some indication of the immense *complexity* of the operation of ideology, which is beyond the scope of this monograph.

8. John Caughie, 'Television and the "Real World": Determination and Representation', unpublished mimeograph, Society for Education in Film and Television, London 1977. On this point, and on issues relevant to this chapter as a whole, see the invaluable essay 'Psychology, Ideology and the Human Subject' by the editorial collective in *Ideology and Consciousness* No. 1, London, May 1977.

9. Poulantzas, op. cit., pp.211-2.

10. P. Francastel, *Etudes de Sociologie de l'Art*, quoted in Stephen Heath, 'Narrative Space', in *Screen* Vol. 17, No. 3, Autumn 1976'. Heath's piece also describes the tyranny of Quattrocento conceptions of space, framing and perspective in the emergence of cinema (and, of course, television).

2 The Practice of History-Writing

This concept [totality], needless to say, is controversial, but then so is the empiricist approach still favoured by the majority of historians in the English-speaking world. (George Lichtheim)

Ideólogy, then, is present at every level of the social formation, the contradictions in that formation (and, more particularly, the concrete conditions of British society), however, ensuring that the dominant ideology—that congenial to the maintenance of the social relations characteristic of the capitalist mode of production—does not go completely unopposed, but that alternative positions are permitted some limited currency. For example, the argument being advanced here will reach whatever public it does by way of a small circulation monograph rather than by, say, a series of leading articles in *The Times* or a one-hour, peak viewing time television programme.

It is characteristic of the dominant ideology that it refines itself out of existence, the *dominant practices* in social institutions and groups becoming naturalised, masking the fact that—far from constituting a natural state of affairs—they too involve theoretical explanations of lived experience. This is as true of the practice of history-writing as of any other (academic) social practice.

It has been an important part of the job of radical workers in all fields to mount a critique of the dominant practices in their areas of concern, to reveal the theoretical and philosophical bases of the dominant practices and to point to their normative implications, a particular necessity in the British context since the dominant practices have traditionally denied that they had theoretical bases or normative implications.

In the field of historiography this task has been carried out most effectively by Gareth Stedman Jones in his essay 'History: the Poverty of Empiricism'.[1] Although he is writing primarily about British historiography, Stedman Jones' critique of the empirical method is valid for the dominant practices in all the social sciences and humanities in Britain. At its crudest the empirical method supposes that any collection of 'facts' will produce their own explanation, leaving out of account the predispositions of the observer categorising them as 'facts' in the first place. Invariably the method draws a distinction between 'facts' and 'experience' on the one hand and 'theory' on the other. It asserts the primacy of the former and, indeed, that the former can exist without the latter.

Stedman Jones describes this as the core of the British historiographical method in the context of history's emergence as an academic discipline in the 1860s and 1870s. The intellectual milieu of Victorian Britain, with the bourgeoisie approaching the zenith of its power,* put on British historiography the firm stamp of liberalism within which men and women were regarded as separate and autonomous beings (and therefore morally responsible for their actions) and within which the optimistic idea of progress was central. Within such a perspective *continuity* was stressed and the central theme of history was seen as the development of civil and religious liberty with the British Constitution and the British Empire as the greatest achievements of 'progress'. (Hegel, interestingly, had traced another, more jagged, line of development to perfection in the Prussian state in which he himself lived.)

While leaning heavily on the positivist notion of 'facts', British historians rejected the positivist impulse to formulate general laws when the facts were gathered, preferring rather to exercise moral judgement. History thus conceived therefore focussed on observable phenomena like the British Constitution and Great Men and was incapable of dealing with non-sensible realities such as class and mode of production. Similarly, the ideas of progress and continuity favoured a kind of linear narrative history whereby 'facts' and 'great men' were assembled in chronological order and suitable moral judgements made about their interrelationship.

However, absolutely at the core of this position was the implicit belief that it was possible to reconstruct the past: in the words of the German historian Ranke, 'simply to show how it really was'. This is a belief that contemporary radical historians would not share. They understand that historiography is not a process of reconstructing the past but of acting upon it — through the traces of it which remain — for the present, of producing a theorised awareness of the past which will shape action in the present. The necessary precondition of this operation is the deconstruction of the concepts which together underpin the dominant mode of historiography.

For instance contemporary radical historians are very much aware that simple temporal associations of events need not imply a structural relationship. Thus, Perry Anderson writes:

> The traditional frameworks of historical writing are either single countries or closed periods. The great majority of qualified research is conducted strictly within national bounds; and where a work exceeds these for an international perspective, it usually takes a delimited epoch as its frontiers. In either case, historical time normally seems to present no problem: whether in 'old-fashioned' narrative studies or 'modern' sociological studies, events or institutions appear to bathe in a more or less continuous and homogeneous temporality. Although all historians are naturally aware that rates of change vary between different layers or sectors of society, convenience and custom usually dictate that the form of a work implies or conveys a chronological monism. That is to say, its materials are treated as if they share a common departure and common conclusion, spanned by a single stretch of time.[2]

*When a dominant class is approaching its zenith or consolidating its power, its intellectuals have no reason to formulate alternative versions of history: the need is felt only in a state of class decline or oppression.

Stedman Jones refers to the traditionally myopic quality of the dominant practice of history-writing in Britain, its failure to take account of the insights of Marxism, psychoanalysis and classical sociology. For a long time the pressure of problems that could not be entirely ignored resulted in these problems being institutionalised within sub-disciplines such as Economic History and Social History, thus leaving the main body of traditional historiography intact.

The awesome power of the dominant ideology is such that even radical historians at a certain stage lived out their revolt against the dominant concern with the historical role of the aristocracy and bourgeoisie by allowing themselves to be 'ghettoised' into Labour History rather than by striking out against the philosophical and methodological inadequacies of bourgeois historiography.

As Stedman Jones points out, there began around 1956 an 'unfreezing of historical debate' in Britain, and Marxian historiography—informed by psychology and sociology—got seriously under way. This had the effect of bringing about a *détente* between bourgeois historiography and bourgeois sociology although the dominant conception of history-writing—and therefore of historical research and history teaching throughout the educational system—remains highly congenial to the (at most, reformist) maintenance of the social relations of production characteristic of advanced capitalism.

It is reasonable to expect, therefore, that television programmes about history will—subject to the qualifications of modernity and fashion—bear many of the features of this dominant conception. We would therefore expect to find many of the following features: a belief in the uniqueness of the event; a belief in the free-will and moral responsibility of individuals; a belief in the role of historical accident (or, more accurately, no search for structural explanations of events); a strong reliance on the role and testimony of individual (particularly 'great') men and women; a strong concern with the nation state both in its political and military dimensions; a belief in progress and in 'chronological monism'.

Conversely, we would expect the relative subordination of features characteristic of the non-dominant (i.e., Marxian) conception of history: the central importance of modes of production; the concept of class; the historical importance of social institutions rather than the actions of individual men and women; the notion of differential temporality in one historical period and the posing of conceptual relationships among events in diverse historical periods (e.g., the English Revolution of 1649, the French Revolution of 1789 and the Russian Revolution of 1917—see Fig. II).

But it must be remembered that we are not here considering conceptions of history in themselves but as they relate to television programmes *about* history. Just as the profession of history-writing evolves its own *practices* which naturalise and therefore mask its relationship with the dominant ideology, so too does the profession of television production evolve *different kinds of practice* which—in their dominant form—are equally congenial to the maintenance of the existing system of power relationships in our society.

The interest lies in the extent to which the dominant practices of history-writing and the dominant practices of television production reinforce each other.

Notes

1. Gareth Stedman Jones, 'History: the Poverty of Empiricism', in Robin Blackburn (ed.), *Ideology in Social Science*, Fontana, London 1972, pp. 96-115. But see also the highly controversial argument, advanced by Barry Hindess and Paul Hirst (in *Pre-Capitalist Modes of Production*, Routledge and Kegan Paul, London 1975) to the effect that Marxism is a theoretical and political practice and, as such, is by its nature incompatible with the essentially empirical practice of history-writing; 'The study of history is not only scientifically but also politically valueless. The object of history, the past, no matter how conceived, cannot affect present conditions. Historical events do not affect present conditions and can have no material effectivity in the present.' This represents, if I understand the argument, the *reductio ad absurdum* of theoreticist Marxism.
2. Perry Anderson, *Lineages of the Absolutist State*, New Left Books, London 1974, pp. 9-10.

3 The Institution of Television and the Idea of Tele-History

There is no such thing as unmanipulated writing, filming or broadcasting. The question is therefore not whether the media are manipulated, but who manipulates them . . . (Hans Magnus Enzensberger)

In the practice of history-writing it is increasingly difficult to avoid consideration of questions of philosophical assumptions and methodology and—in the academic milieu at any rate—there is the time if not always the will for recurrent consideration of these issues.

The milieu of television production is altogether different, the fact of having to work to tight schedules and the generally accepted view of the transience and disposability of the object being produced militating against such self-reflection.

The day-to-day working principles whereby television programmes are put together have been so usefully summarised by Stuart Hall that one can do no better than quote him at length:

> On the whole they seem to amount to recipes and tags which can be integrated within the notion of 'what constitutes *good* television'. Good television captures the on-going form of real life; it is close to actuality. Good television is smoothly edited, chaired and presented—it offers a polished professional product, largely in terms of the assembly of material, the smooth management of the transitions. It is 'good television' if there are no breaks or discontinuities. Despite recent innovations, it is still the case that, by and large, in 'good' professional television, the technical nature of the medium is suppressed: any reminders that there are cameras, cameramen, sound booms, interviewers, commentators, technicians, secretaries, script-girls, location- or studio-managers, producers, directors, telecine recordists, and so on, invisibly intervening as a collective production unit between 'reality' and the viewer, destroys the illusion of immediacy and transparency. Some of the more 'creative' programmes—especially in the comedy and discussion area—do make use of the television 'hardware' as part of the set; but standard television production, which aims at the polished professional product, still characteristically suppresses the technical component.
>
> All television must be 'rapid'—it is not a medium which must be permitted to stray from its point, desert or digress from its original programme conceptions, turn up unexpected materials, etc. 'Good' television talk is brief, concise, makes one or two points only in a clear and simple manner. Television is not a medium for complexly structured

or nuanced argument or exposition. 'Good' television conversation is well-bred, allows a balance of points of view, never disrupts the even tenor by shows of anger, deep commitment, flights of rhetoric, symbolic or metaphoric statement. 'Good' television is, essentially, majority television: that is for an audience composed not—as it almost certainly is—of different groups, with cross-cutting minority interests, but as a large, undifferentiated, homogeneous mass. Thus 'good' television must be either plain, simple and straight, *or* it requires the mediation of the explainer/guide/moderator, who 'stands in' for the absent audience and makes the complicated plain, simple and straight.

There are, of course, two variants of this gospel: the demotic variant, which pretends that television *is* exactly real life, where 'the people' are, 'what the people want'; and the paternalistic variant, which pretends that the medium itself must continually translate complex realities into the simple terms which 'the man in the street' can comprehend. 'Good' television visualises whenever it can, never uses a word when it can supplant it with an image or an illustration, and is constantly beset by people who do not understand its visual mysteries and who insist that television should sound intelligent as well as look professional. 'Good' television is visually dramatic, the pictures are full of incident. 'Bad' television is static, talking heads, long camera takes, pictures which do not 'move' . . .[1]

As will be demonstrated presently, these principles are subscribed to as heartily by producers of 'factual' programmes about history as by producers of historical drama and light entertainment shows.

There is considerable debate even within the bourgeois historiographical consensus about the inadequacy of 'factual' tele-history programmes, and a particularly interesting debate about the nature of film as evidence.[2]

As is to be expected, however, the reservations are for the most part formulated from a different perspective from this monograph, often as an attack on the bureaucratic/financial pressures surrounding the historian collaborating on a tele-history project rather than as dissent from the dominant practices of television production. Thus, Donald Watt—a historian who has been involved in tele-history—attacks the populist prejudices of television administrators; the implications of the heavy costs of historical series (international co-production deals, short preparation and shooting schedules, the 'dissipation of know-how' by taking on researchers on short-term contracts and having to take on totally new teams for subsequent series, the ratings battles between channels); the aspiration to entertainment; and—as a somewhat separate (but entirely accurate) point—the lack of serious television criticism in general; but particularly in response to the tele-history series.[3]

In so far as tele-history has an explicit theory it has been adumbrated by Jerry Kuehl, himself producer of Thames Television's *World at War* and *Destination America*. His views are remarkable for the extent to which they reflect the general consensus regarding 'good' television outlined above and for their seeing the consensus as 'iron laws' of television production rather than as socio-aesthetic conventions formulated within the dominant ideology and designed to give it the least inhibited passage. Thus:

> The first thing the good [television producers] learn is that what they make are television programmes; that is to say works which should follow the rules of *television*—which are not at all the same as those which govern the production of learned articles, or indeed, purely literary works of any sort. The second thing they learn is that their audience is a mass audience . . .[4]

and

> One characteristic of television as a communication medium is that it offers its audience virtually no time for reflection. It is a sequential medium, so to say, in which episode follows on episode, without respite. This clearly means that the medium is ideally suited to telling stories and anecdotes, creating atmosphere and mood, giving diffuse impressions. It does not lend itself easily to the detailed analysis of complex events; it is difficult to use it to relate coherently complicated narrative histories, and it is quite hopeless at portraying abstract ideas.[5]

Bearing in mind the expectations already expressed above (chapter 2) of what the dominant conception of history would encourage and discourage on television, this list — pro-phenomenal event and anti-structural analysis[6] — reads like a charter for bourgeois historiography.

By posing the notion of the *rules* of television, Kuehl rules out of court the possibility of alternative structures and aesthetic practices which *could* accommodate some of the things he suggests television cannot intrinsically do. The fact that television audiences are offered no time for reflection (or direct feedback) is not *inherent* in the medium itself, it is the consequence of a cluster of factors the most important of which are the duopolistic control of transmission: the conception of 'the evening's entertainment'—as in Raymond Williams' term 'flow';[7] and the constituting of the audience as an undifferentiated and passive mass.

Certainly there are powerful conventional as well as directly ideological pressures against television handling 'the detailed analysis of complex events' and 'portraying abstract ideas' but these too are not rendered impossible by *intrinsic* features of the medium but by received practice and aesthetic convention whereby, in film and television, experiential phenomena are given primacy over equally real, but not directly observable, deep structures. That 'visual' media can handle 'the detailed analysis of complex events' and 'abstract ideas' is amply testified to by the work of Dusan Makaveyev and Jean-Luc Godard, among others.

Stuart Hall remarked on the primacy of the image over the word in the received wisdom of television production. This has positively alarming consequences in historical documentary on television as Jerry Kuehl, discussing the role of commentary, and apparently unperturbed by the situation he describes, writes:

. . . the gaps in commentary may be dictated, not by the writer's conscious decision . . . but by what is or is not available on film. An example again: relations between Church and State were very important to the leaders of the Third Reich, and, it goes without saying, to ordinary Germans too. But very little film was ever made which even showed National Socialist leaders and churchmen together, let alone doing anything significant. *So considerations of Church and State were virtually omitted from our films on Nazi Germany — and from our commentary.* [8] [My italics]

Such is the tyranny of the moving image in tele-history that the existence or non-existence of a piece of film may determine whether or not a particular historical point will be made. The alternative of working out different aesthetic strategies — e.g., direct address, diagrams or animated drawings, photographing documents, etc. — is inadmissible primarily because they would draw much more attention to the *constructed* quality of the programme and at the same time diminish the 'authority' of actuality footage, an authority, it will be remembered, which is not natural, but which flows from the primacy given the phenomenal world as perceived by the senses within the empirical philosophical tradition.

The collection of tags and maxims Stuart Hall describes above constitute the observable features of the *institution* television, and the disagreements referred to between historians and television producers constitute the interface between the *institution* television and the *institution* historiography.

But what is this *institution* television and what is its ideological import as an *institution* as opposed to the ideological import of its discrete signified messages? This is a problem which has not gone unnoticed but which has largely gone *untheorised*. When John Birt and Peter Jay talk of the 'bias against understanding' in television coverage of current affairs as expressed in, for example, the over-use of dramatic war footage and the poverty of analysis, and when Raymond Williams says that 'one of the innovating forms of television is television itself', they are broaching this problem, as are the commentators who point to the presence of the television set continuously in the corner of the room and our surrender to the carefully constructed inducements to leave it switched on.

An attempt has been made to *theorise*, as opposed simply to assert, the ideological import of the institution television by Stephen Heath and Gillian Skirrow,[9] who criticise the communication model of television as sender/code/ message/receiver which is very widely held both by academics and practitioners — a model which carries within it the possibility of characterising television as an ideologically neutral institution through which pass, undefiled, the messages of the broadcasters to their audiences. Heath and Skirrow force attention upon the institution television as constructor of a reality which is distinct from the social world inhabited by both broadcasters and audiences and which is subject to its own principles of construction that have to do primarily with the institution television rather than with the ostensible topics of the programmes. Describing the central function of the institution television as 'the occupation of the viewer as subject in a permanent arena of communicationality', Heath

14

and Skirrow demonstrate how an episode of *World in Action* (Granada), ostensibly about truancy, is really about the engagement of the viewer in dramatic narratives constructed round the opposition yesterday/today and round certain stereotypes of young people.

While much of this monograph is concerned with the ideological import of specific television messages, it concurs with the position adumbrated by Heath and Skirrow that it is the *institution* television which is of primary ideological significance rather than the points of view of particular broadcasters and much of the detailed discussion herein will point to the determinate nature of television *per se* rather than historiography in what actually happens on the screen.

However, relatively autonomous as the institution television is, the argument of this monograph thus far has been that *superstructural* activity such as the writing of history and the production of television programmes are not in the last analysis diverse, wholly autonomous and unconnected activities but take their character from the system of production relationships in the social formation they inhabit. To be sure, the activities of history-writing and television production (and every other *superstructural* activity) evolve their own particular *practices* which mask their relationship with each other and with the economic base, but since the central function of superstructural activity is the operation of *ideology*, the most important question to be asked of any *practice* is whether it is congenial or uncongenial to the maintenance of the existing relations of production in our society.

It has been suggested that the *dominant* practices in British historiography and in British television production—through their subscription to the empirical philosophical position—are highly congenial to the maintenance of the socio-economic *status quo* and that, in so far as the practice of tele-history has an explicit theory, it too is similarly congenial. These arguments must now be expanded and their validity assessed in relation to concrete tele-history and historical drama programmes.

Notes

1. Stuart Hall, 'Television and Culture' in *Sight and Sound*, London Autumn 1976, pp. 246-252.
2. See Paul Smith (ed.), *The Historian and Film*, Cambridge University Press, Cambridge 1976.
3. Donald Watt, 'History on the Public Screen I' in Paul Smith (ed.), op. cit., pp. 169-176.
4. Jerry Kuehl, 'History on the Public Screen II', in Paul Smith (ed.), op. cit., p. 178.
5. Ibid., pp. 178-9.
6. A directly analogous criticism has been made, with regard to news coverage on British television, by John Birt and Peter Jay in *The Times*.
7. Raymond Williams, *Television: Technology and Cultural Form*, Fontana, London 1974, pp. 86-118.
8. Jerry Kuehl, op. cit., p. 179.
9. Stephen Heath and Gillian Skirrow, 'Television, a World in Action', *Screen*, Vol. 18, No. 2, Summer 1977.

4 The Historical Role and Testimony of the Individual

It is impossible to know anything about men except on the absolute precondition that the philosophical (theoretical) myth of man is reduced to ashes. (Louis Althusser)

As has been suggested, the constituting of men and women as separate and autonomous beings, as *individuals*, was a central element—at the *philosophical* and *ideological* level—in the rise to dominance of the bourgeoisie. It is to be expected, therefore, that the category of the *individual* will be a central structuring category in all superstructural activity in bourgeois societies. In the area of 'fictional' and 'factual' television programmes about history, this is overwhelmingly so.

With regard to historical drama, i.e., those programmes which are classed as drama but which involve actual historical personages, there are few programmes not so structured. This is reflected in their titles: *Edward the Seventh* (ATV), *The Six Wives of Henry the Eighth* (BBC), *Jenny* (Thames): the story of Winston Churchill's mother, *Eleanor Marx* (BBC), and so on. Certain other historical drama series, while not overtly biographical in their titles, nevertheless have been structured, in their weekly episodes, round individual historical personages: e.g. *The Velvet Glove* (BBC), which consisted of programmes devoted to the lives of Edith Cavell, Marie Stopes, Mary Baker Eddy, Elizabeth Fry, Annie Besant, and Lillian Baylis.

The effect—whether conscious or otherwise—of structuring historical drama round the lives of (usually famous) individual men and women is to suggest that history is made primarily by the individual interventions of men and women acting as free agents, rather than by the complex interplay of *classes*, *institutions* and *modes of production*. This is not to argue that the contributions of individual men and women to the making of history can be overlooked, simply that they must be reconceptualised and seen in relationship with other factors.[1]

The cry will go up that, of course, television historical dramas are structured round the lives of individuals, because 'drama is about people'. Unquestionably, television drama has got to be meaningful in relation to the lived experience (including the need for the play of imagination) of the audience, but there is no *intrinsic* reason why it must be primarily concerned with the lives of individuals or with inter-personal relationships. It is possible to conceive of other kinds of aesthetic strategy which might foreground *groups, classes* and *institutions*. (The BBC television adaptation of *The Cheviot, The Stag and the Black, Black Oil*, discussed in chapter 10, is partly successful in this regard.) The argument being advanced here is that television (historical) drama is overwhelmingly structured round the category of the individual for *ideological* reasons. Not that there is necessarily a conspiracy to exclude other forms of historical drama (although such a conspiracy would, indeed, be one of the symptoms—not the *cause*—of bourgeois dominance in television historical drama), but that the category of the *individual* is regarded as a *natural* structuring category in the milieu of television (historical) drama.

The available evidence suggests that—as in so many other fields—television personnel of a radical persuasion are living out their working lives either willingly or unwillingly imprisoned within the categories of the dominant conception of (historical) drama. Thus, even the extremely important and interesting *Days of Hope* (BBC) was structured round the lives of three working-class people—the television drama equivalent of radical historians having been ghettoised in the past into Labour History before they broke out and challenged the rules of the history-writing game.

The structuring of 'factual' television programmes about history round the category of the individual is perhaps less explicit than in 'fictional' programmes but is nevertheless very widespread and takes several forms. The most usual practice is to have an institutional, periodic, or geographical title *(The British Empire*, BBC; *Civilisation*, BBC; *The Great War*, BBC; *Destination America*, Thames, etc.) and a *de facto* structuring round the category of the individual.

It is instructive to look closely at the form of two episodes in the series *The British Empire* entitled 'The Sugar Slaves' and 'In Darkest Africa', the one dealing with the British connection with the West Indies, the other with the exploration of Africa. The point of looking at *two* episodes rather than *one*— and two episodes in which key figures like the writer were different—is to argue that the conception of tele-history represented by the series transcends the attitudes of this or that television writer, director or producer and is embodied in the broadcasting institutions themselves as part of the superstructure of a particular bourgeois society.

The extremely important questions of the coding of *The British Empire*'s opening imagery, theme music and the typography of its title on the screen; of the coding implied by the choice of Robert Hardy as narrator (see chapter 5); and of the coding implied by each episode beginning (as do so many television drama programmes) with a 'hook' in the form of a dramatic or arresting incident; all require discussion but are somewhat separate from the present discussion of how the category of the individual becomes hegemonic even when the programmes are ostensibly dealing with other questions.

In the two episodes of *The British Empire* under scrutiny, the category of the individual is reinstated by tracing the development of the geographical/historical areas through the lives of individual men. Thus, early in 'The Sugar Slaves' there is an account of the 'discovery' (in bourgeois accounts of the history of the Americas, places are 'discovered' when the first European lands there, irrespective of the length of tenure of the indigenous peoples) of the island of St Christopher in 1642 in relation to the life and career of one Thomas Warner and the subsequent wresting of colonial control of the Caribbean from Spain and Portugal is partly related in personalist terms—'. . . Warner took Nevis, Antigua, Montserrat . . ., the Dutch took . . . and the French took . . .'

Similarly, quite early in 'In Darkest Africa', discussion of the British attempt to enforce abolition of the West African slave trade has, as the images behind the commentary, a set of watercolours depicting naval engagements off the African coast. These are revealed to be by the hand of the commander of a naval squadron involved, Henry Need, and give way to his watercolours of the flora, fauna and people of West Africa. The commentary goes on to talk about the passion for exploration and scientific knowledge of the time, which—by way of an image of a museum drawer full of butterflies, the collection of Joseph Banks— leads into an account of Banks' life and his contribution to the opening up of Africa.

Both episodes continue to proceed by way of the contributions of particular key or typical individuals: Mungo Park, the Landers, David Livingstone, etc., in the case of 'In Darkest Africa'; and John Tharp, Prince William and Horatio Nelson in the case of 'The Sugar Slaves'. To be fair, notions such as class, profit, and exploitation are by no means absent from the discourse of these pro- grammes—indeed, *The British Empire* series was attacked by many people, including some professional historians, on account of its 'facile, anti-imperial bias'. What is at issue is what is *foregrounded* in the coding of these programmes and what this implies for the conception of history which underpins them and, by extension, for their relationship to bourgeois ideology.

As has been argued in other monographs in this series,[2] all television—whether 'factual' or 'fictional'—in a bourgeois society aspires to the condition of enter- tainment and partakes, thereby, of proven aesthetic strategies towards achieving this end, towards delivering the kinds of audience satisfactions commensurate with conceptions such as 'the evening's entertainment' and television as 'flow'. The hegemonic category in television is, of course, one which partakes of the dominant structuring category of the individual—that is, the *star*. The star and his/her extensions, family, car and—pre-eminently—*home* are also the focus of discussion in television's back-up documentation, *Radio Times* and *TV Times*.

There is a somewhat striking example of this in the two *The British Empire* episodes under discussion. Making the legitimate point of the penetration into British aristocratic and political life of the West Indian plantation families, 'The Sugar Slaves' spends quite a long time (relative to its overall length) giving an account of Tharp's English country house and those of other planta- tion owners. There is a directly analogous scene in 'In Darkest Africa' with a long (again, in relation to the overall length of the programme) examination of the English country house of Lord Charles Somerset, Governor of the Cape Colony.* If we see 'the evening's entertainment'—including 'factual' as well as 'fictional' programmes—and *TV Times* as inhabiting the same ideological

*However, the most bizarre and, for the present writer, the most suffocating manifestation of the festishisation of the home is in *The Ascent of Man* (BBC). The ostentatious display of property which is at the root of this motif is joined with a particular ideology of that series within which Bronowski's California home is transmuted into a secular and humanist Sistine Chapel where the Great Books and Great Art of Mankind become, under the Doctor's priestly hands, objects in the mystical cult of Learning.

18

universe, then the connection (between the aesthetic strategy of filming the English country houses of the aristocracy for programmes about history and the long-running feature in *TV Times* within which 'stars'—both of 'factual' and 'fictional' television—are discussed in relationship to their homes) may seem rather less absurd than if posed casually.

The centrality of the contribution of 'great men' to the making of history may be found cheek by jowl in television programmes about history with another assertion of the primacy of the individual as a category of historical discourse— the *testimony of individual men and women.* This latter is the central strategy of the series *Destination America,* each programme of which consists of inter- views with descendants of the various national and ethnic groups who left Europe to make their homes in America, intercut with actuality film relevant to the immigrant experience. Its impulse is thus an extremely generous and progressive one, a kind of television equivalent of the 'history from below' movement. However, the argument of this monograph is that, by the nature of its structuring categories, *Destination America* remains locked within the bourgeois conception of history.

For example, 'On a Clear Day You Could See Boston' deals with the migration of the Irish to, and their settlement in, America. Its structure—predictably, since it inhabits the same ideological universe—is not dissimilar to the episodes of *The British Empire* discussed above, even to the extent of having a 'hook' in the form of a pre-credit sequence (the formal fact of the existence of the pre-credit sequence being a further ideological identifying feature) of John F. Kennedy's visit to Galway. This sequence fulfils the double purpose of dramatic 'hook' and of advancing the thesis that history is primarily about individual, particularly 'great', men. The ideological implications of *Destination America*'s credit sequence are important, though it would be inappropriate to go into detail here. Suffice it to say that the Dvorakian music, the dissolves from one immigrant face to another and the turning of the Statue of Liberty from darkness into light appear to give uninhibited passage to the ideology of America as 'God's crucible'.

It is no part of the argument of this monograph that programmes such as *Destination America* are 'bad' or inaccurate. The particular episode dealing with the Irish gives an immense amount of information about the Irish in relation to America: the reasons for, and rigours of, the crossing; where Irish communities settled; the centrality of Boston in the Irish experience; the anti-Irish prejudice of the 'native' Bostonians; the importance of the Irish in the construction industries and in politics; and the relationship of the American Irish to present- day events in Ireland. The argument is about the nature of the categories used to structure this information and their ideological implications. The fact that the main recurrent structural element is the on-camera interview with descendants of the migrants in itself asserts the explanatory force of individual experience and testimony *vis-à-vis* history. As it happens, the content of that testimony in many cases surrenders to rhetorical and mythical explanations of the Irish experience as do some of the interventions of that invisible but omnipresent force in tele-history (as in a great deal of other film and television 'documentary'), *the narrator.*

Notes

1. This problem is recognised and debated among Marxists. See, for example, George Novack, 'From Lenin to Castro: The Role of the Individual in History Making' in *Understanding History*, New York 1972, and E. H. Carr, 'Society and the Individual' in *What Is History?*, Penguin, Harmondsworth 1961. However, the formulation of the relationship 'individual' and 'society' as being between two separate and autonomous entities is inadequate. For a more genuinely materialist formulation see 'Psychology, Ideology and the Human Subject' in *Ideology and Consciousness* No. 1, May 1977: '. . . human subjects are "always-ready social", that is to say that they only exist as human subjects in and through a set of relations that pre-exist any "individual". Indeed, not only the "content" but also the "form" of what we take in common sense to be our individuality is itself a function of these relations. And these relations are indeed a "set", that is to say they are interconnected and inter-dependent. Individual relations cannot be thought of in isolation, or reified (as, for example, in "the necessity of the mother-child bond" or "the structure of the family") for they only exist in relation to other relations and, ultimately, to the relations of production and reproduction.'
2. For example, Richard Collins, *Television News*, BFI, London 1976, and Edward Buscombe (ed.), *Football on Television*, BFI, London 1975.

5 The Narrator as Guarantor of Truth

. . . the historian or relater of things important to mankind must, whoever he be, approve himself many ways to us . . . ere we are bound to take anything on his authority. (3rd Earl of Shaftesbury)

There is a justly famous sequence in *Letter from Siberia*, a film by the distinguished French documentarist Chris Marker, in which the series of images just seen by the audience is replayed twice with different commentaries which totally reverse the ideological meaning of the sequence as first shown. This illustrates starkly and directly what most of us know but constantly have to dredge up to the surface of our consciousness, i.e. that while filmed records of events are by no means ideologically neutral,* they nevertheless, in themselves, lack total explanatory force and require supplementing by other codes—usually musical or verbal or, most particularly, montage codes—before this explanatory force can *begin* to come into play.

So omnipresent is the phenomenon of narration and the (most usually disembodied) narrator, so *naturalised* has the process become, that considerable effort is required to distance oneself from it and interrogate its ideological import. This is true of documentary film practice in general and, within this, of 'factual' programmes about history. So prevalent is the fact of narration and the conception of history as narrative that the highly controversial issue[1] of the *necessity* of the relationship between narrative/narration and historiography has been (particularly with regard to tele-history) repressed.

It will readily be conceded that in 'real' life people's voices and verbal language use (like their dress and physical mannerisms) are coded, giving information of a complex kind. This, naturally, can become part of the film or television message (although it should be noted that the information 'urban, working-class Scot' passed by linguistic codes, while related to, is not the same as, the information 'actor *signifying* urban, working-class Scot'). This fact therefore should provoke reflection about the choices made as to which figures act as narrators in television programmes about history and what the ideological implications of these choices might be.

*This is a difficult and complex question but some of the issues which have to be thought about are whether the construction of lenses so as to give monocular, Renaissance perspective constitutes an ideological choice and whether the use of certain film stocks rather than others and certain filming practices (e.g. hand-held camera) constitute acts of coding. For instance, the exterior sequence in Kubrick's *Dr Strangelove* (1963) showing the infantry attack on General Ripper's base, seems to have been coded (through the particular film stock, lack of 'composition', and jerky camera movements) to evoke in the audience the sense of watching a Second World War newsreel, a strategy also used in Godard's *Les Carabiniers* (1963).

Right from the start of tele-history, narration has been the preserve primarily of the *actor* (significantly not the *actress* except in manifestly 'feminist' programmes such as *Women at War*) when the narration has been off-camera. When the narrator has been for a large part of the time on-camera (as in *Civilisation, The Ascent of Man, America, The Age of Uncertainty,* etc.) then a somewhat different coding operation is at work, as will be discussed presently. The argument of this monograph is that the central ideological function of the narration is to confer *authority* on, and to elide *contradictions* in, the discourse. In a *patriarchal* society the narrator must therefore be a man and in a *bourgeois* society his voice must be one which signifies bourgeois authority. While few choices of narrator carry such flagrant ideological overtones as the use in the American series *Victory at Sea* of Richard Burton in his Churchillian dimension, the narration of tele-history has operated firmly, if not exclusively, within the linguistic parameters of the BBC and the London stage at an earlier point of their development (i.e., prior to the mid-fifties, when regional and non-upper class accents increasingly penetrated the 'serious' productions of these institutions). This is exemplified by the choice as narrator* of Michael Redgrave (*The Great War*, BBC), Laurence Olivier (*World at War*, Thames), Iain Cuthbertson (*Destination America*, Thames)—using of course his Standard English voice rather than his *Sutherland's Law* (BBC) voice—and Robert Hardy (*The British Empire*, BBC). To be highly speculative, it is interesting to ponder what additional reasons might exist—apart from linguistic coding— as significatory of the ideological projects of these programmes. Could perhaps the Redgrave and Olivier knighthoods (and, in the latter's case, a peerage) have been relevant; could the choice of an actor associated in the public mind with regional roles be regarded as more fitting to the demotic aspirations of *Destination America*; and could the well-known association of Robert Hardy with English tradition (e.g., his narrating the series *Heritage*, BBC, on English pageantry and his book and television programme on the English long-bow) have been an added *cachet*?

Unquestionably, narration in tele-history has strong *denotative* features, i.e., the *matter* of what is said is crucially important: 'hard' information is given about which armies took part in particular battles, where the battles occurred, what the outcome was; how many of a particular nationality came to America and when, where they settled and what sort of space was allowed them, and so on. However, closely allied to the (subconscious?) class choices of narrator referred to above are the *connotative* aspects of narration, the *rhetoric* of narration. This is, of course, quite difficult to separate out from the rhetoric of other codes (pictorial, musical, sound) working simultaneously with it, but it might be argued that the rhetoric of narration in tele-history veers towards

---

*It is of the essence of ideology that its operations are masked from those within it (see chapter 1 above). I would expect the programme makers to claim (nor would I disbelieve them) that the choice of narrator is made on the grounds of lucidity of speech and capacity for 'drama'. The latter reason would throw interesting light on the running argument of this monograph about the aspiration of *all* television—'factual' as well as 'fictional'—towards the condition of entertainment.

that of more heightened, stylised forms of verbal discourse such as blank verse and other *dramatic* forms. This would, of course, be consistent both with a view (whether conscious or otherwise) of *all* television as 'entertainment' and, at the same time, the apparently paradoxical impulse to signal the *difference* of tele-history from the comedy show which precedes it and the police series which comes after it, to signal its *seriousness.*

It is necessary, in its aspiration to drama, that the narration of tele-history be uneven, have troughs and *crescendi,* the latter being associated with the beginnings and endings of programmes, with structural features of the institution television such as commercial breaks, and with the impulses of the narrative itself (e.g., the points of transition from stasis to action in a series about war). Needless to say, the *crescendi* of narration are supported by *crescendi* in the other codes, most notably those of music and cutting.

Take the complex codification of the opening of any episode from *The Great War*. The credit sequence consists basically of three extremely dramatic photographs from the First World War: a soldier standing beside a cross on the skyline; a half-decayed body in a trench; and a British soldier staring, as if shell-shocked, at the camera. The credit sequence is paced by Wilfrid Josephs' bleak and evocative musical score and the photographs are connected by disturbing camera movements downwards and sideways. The sequence is clearly designed to evoke the sense of horror associated with certain aspects of the First World War which have registered in popular memory: its trench warfare; its dashed hopes; and its casualty lists in millions. However, the view of the war which the sequence evokes is of a timeless and metaphysical category prised free from history rather than as a conflagration which broke out as a result of the convergence of quite concrete historical forces. Given a recurrent opening motif of such dramatic force, the rhetoric of the narration could not then relapse into *pure* 'hard' information-giving or analysis: dramatic necessity requires that it aspire, from time to time at least, to the same register as the credit sequence. Thus, the episode entitled 'This Business May Last a Long Time'—concerned with the change from classical to trench warfare—begins with shots of German soldiers in repose accompanied by the cadenced narration of Sir Michael Redgrave:

> The pendulum of war had come to rest; the armies halted. Round the camp fires men were too weary to talk much, but they *could* wonder which way would they march tomorrow . . .

However, important as the 'hard' information-giving and the rhetorical features of narration are, far and away its most important function is as organiser of the other discourses constituting the programme and as guarantor of its 'truth'.

This feature of narration has been well described, with regard to the nineteenth century novel and the classical narrative fiction film deriving so largely from it, by Colin MacCabe. The 'classic realist text', as MacCabe calls it, is characterised by:

> a hierarchy amongst the discourses which compose the text and this

23

hierarchy is defined in terms of an empirical notion of truth. Perhaps the easiest way to understand this is through a reflection on the use of inverted commas within the classic realist novel. While those sections in the text which are contained in inverted commas may cause a certain difficulty for the reader — a certain confusion *vis-à-vis* what really is the case—this difficulty is abolished by the unspoken (or more accurately the unwritten) prose that surrounds them . . . Whereas other discourses within the text are considered as material which are open to re-interpretation, the narrative discourse simply allows reality to appear and denies its own status as articulation . . .[2]

The analogous transparency of the narration in tele-history programmes gives it enormous ideological force, particularly since the voice speaking the narration is—as has been suggested above—almost invariably the voice of bourgeois authority.

The above quotation from MacCabe might have been written with *The Great War* in mind, for the latter is constituted largely by attributed texts with 'quotation marks' round them: 'As General X's order of the day said . . .'; 'Colonel Y, writing home from the front, said . . .'; 'As an editorial in the *Frankfurter-Zeitung* said . . .'; and so on. Also in 'inverted commas', so to speak, are the on-screen recollections of First World War veterans which form a recurrent discourse in the series. Probably the only two discourses without 'inverted commas' are the actuality footage round which the series is built and the narration. Both are presented as transparent, unarticulated, self-evidently 'true' and are presented as such for separate ideological reasons which derive from the same philosophical basis (see chapters 1, 2 and 3 above).

It is worthwhile recalling the remarks of Stuart Hall quoted at some length in chapter 3, particularly his suggestion that, in the received wisdom of television production, ' "good television" must be either plain, simple and straight, *or* it requires the mediation of the explainer, guide, moderator, who "stands in" for the absent audience and makes the complicated plain, simple and straight'. In this latter category Hall had primarily in mind professional 'link-men' such as Robin Day and Michael Barratt, but the two types he poses fit the types of tele-history. If series such as *World at War*, *Destination America* and *The British Empire* constitute the former, *Civilisation*, *The Ascent of Man* and *The Age of Uncertainty* constitute the latter. In one the authority is covert, in the other — by reason of the academic reputations of the narrators — it is overt, indeed *flaunted*. In each case the narrator guarantees *Truth*.

MacCabe makes the important argument that 'the classic realist text cannot deal with the real in its contradictions and . . . in the same movement it fixes the subject [i.e., the spectator] in a point of view from which everything becomes obvious'.[3] He then goes on:

There is, however, a level of contradiction into which the classic realist text can enter. This is the contradiction between the dominant discourse of the text and the dominant ideological discourses of the time. Thus a classic realist text in which a strike is represented as a just struggle in which oppressed workers attempt to gain some of their rightful wealth would be in contradiction with certain contemporary ideological discourses and as such might be classified as progressive. It is here that subject matter enters into the argument and where we can find the justification for Marx and Engels' praise of Balzac and Lenin's texts on the revolutionary force of Tolstoy's texts which ushered the Russian peasant onto the stage of history. Within contemporary films one can think of the films of Costa-Gavras [e.g., *Z, State of Siege*] or such television documentaries as *Cathy Come Home* [BBC]. What is, however, still impossible for the classic realist text is to offer any perspectives for struggle due to its inability to investigate contradiction. It is thus not surprising that these films tend either to be linked to a social-democratic conception of progress — if we reveal injustices then they will go away — or certain *ouvrieriste* tendencies which tend to see the working-class, outside any dialectical movement, as the simple possessors of truth.[4]

Written primarily with narrative fiction in mind, this quotation is consequently entirely applicable to television historical drama with its discourse organised by the unseen writer and television director whereby, as MacCabe has put it, 'its end is guaranteed from its beginning and it is this certainty which enables the reader to place him or herself in a position of unity from which the material is dominated'.[5] But patently, it is also applicable to 'factual' television programmes about history where everything is filtered either covertly, through the absent writer and off-screen narrator, or flamboyantly through the sensibility of the on-screen writer/narrator.

The only 'progressive realist text' I am aware of in the area of historical drama on British television is *Days of Hope*, although its massive retention of the features of realist drama is, at the very least, problematic.[6] Such texts exist in other areas of television, e.g., *World in Action*[7] and the John Pilger documentaries which ATV nervously announced as constituting 'a personal view' and hastily followed with a programme by Auberon Waugh, certain of whose views are pronouncedly right-wing, for the sake of 'balance and impartiality'. But although the former are progressive, they are nevertheless still 'realist' and as such — in MacCabe's terms — they filter contradictions through a particular point of view and resolve them for the audience.

If the 'progressive realist text' is thin on the ground in the area of historical drama and 'factual' programmes about history, the 'revolutionary text' as canvassed by MacCabe — that text in which contradictions are not resolved for the spectator but which, through its critique or total abandonment of 'the hierarchy of discourses', leaves the spectator with a deal of work to do to resolve and act out the contradictions — that text is *absent* from the television screen, but some progressive answers to the central problem it poses are discussed in chapter 10 of this monograph.

25

Notes

1. See the journal *History and Theory* intermittently over the past fifteen years, but particularly Vols VI, VIII, X and XV; and—more recently—Leon J. Goldstein, *Historical Knowing*, University of Texas Press, 1976.
2. Colin MacCabe, 'Realism and the Cinema: Notes on some Brechtian Theses' in *Screen*, Summer 1974, pp. 8-9.
3. Ibid., p.16.
4. Ibid.
5. Colin MacCabe, 'Days of Hope' in *Screen*, Spring 1976, p.100.
6. See the argument on *Days of Hope* between the present writer and Colin MacCabe in *Screen*, Winter 1975/6 and Spring 1976; and (also relevant to the issue of Realism) the generous text by Raymond Williams—'A Lecture on Realism'—in *Screen*, Spring 1977.
7. See, for example, the article by Stephen Heath and Gillian Skirrow, op. cit.

6 The Legacy of Positivism and the Authority of the 'Real'

Realism is an issue not only for literature: it is a major political, philosophical and practical issue and must be handled and explained as such — as a matter of general human interest. (Bertolt Brecht)

Much has been said in the preceding chapter about the elision of contradictions in tele-history and the authority conferred on programmes through the ideological mechanism of the narrator. The effect of this is that tele-history presents itself as unproblematic. To be sure, this or that essay in tele-history will be subtitled 'a personal history' but the mechanisms deployed in the 'writing' have the effect of proposing the removal of tele-history from the realm of opinion (and, therefore, of ideology) into that of *Fact*.

The confident, developmental surge of series such as *The Great War* and *The Fight Against Slavery* is perhaps to be expected, but it is supremely ironic that much of a series entitled *The Age of Uncertainty* should be cast in the positivistic form of nineteenth century historiography. As E.H. Carr has written:

> First ascertain the facts, said the Positivists, then draw your conclusions from them. In Great Britain, this view of history fitted in perfectly with the empiricist tradition which was the dominant strain in British philosophy from Locke to Bertrand Russell. The empirical theory of knowledge presupposes a complete separation between subject and object. Facts, like sense impressions, impinge on the observer from outside and are independent of his consciousness. The process of reception is passive: having received the data, he then acts on them. The Oxford Shorter English Dictionary, a useful but tendentious work, clearly marks the separation of the two processes by defining a fact as 'a datum of experience as distinct from conclusions'. This is what may be called the commonsense view of history. History consists of a corpus of ascertained facts. The facts are available to the historian in documents, inscriptions and so on, like fish on the fishmonger's slab. The historian collects them, takes them home, and cooks and serves them in whatever style appeals to him. [1]

Just as there has been considerable debate among historians and philosophers of history about the issue of narrativity (see chapter 5 above) [2] so too there has been considerable debate about what constitutes historical fact. This awareness — and the impulse it contains to foreground questions of historical method — is repressed in tele-history so that the unspoken models of series such as *The Great War*, *Civilisation* and *The Ascent of Man* are the great rhetorical works of an epoch (marking, as it did, the rise to impregnability of the bourgeoisie) which required no space for reflection upon how facts are constituted as such, works such as those by Gibbon and Macaulay.

As E.H. Carr has pointed out, not only do bourgeois historians fetishise 'facts', they also fetishise visible evidence of 'facts' in the form of documents — decrees, treaties, rent-rolls, blue-books, official correspondence, private letters and diaries. Such evidence is visible and tangible as befits the Anglo-American philosophic disposition.

Operating, as it does, within an ideological framework inherited from nineteenth century positivism, tele-history, unsurprisingly, has its own analogous fetish. Just as the bourgeois historians of the last century pinned their faith on *documents* as — in E.H. Carr's neat phrase 'the Ark of the Covenant in the temple of facts' — so do the tele-historians of the twentieth century pin their faith on actuality film or, when that is not available, on the next best thing, reconstruction film, and the return to the 'actual scene' where particular historical events occurred.

It has been demonstrated (see p. 14 above) that the existence or non-existence of footage may determine whether or not a particular historical area may be dealt with, in other words the *determinant of television* rather than that of *historical knowing* comes into play. This is the reverse of the coin whereby the *existence* of actuality film will virtually guarantee its use, irrespective of its relevance to historical argument. This is a by-product of the philosophical belief in the authority of the *Real*, the real being conceived as the phenomenal world.

It is a useful exercise, when looking at an episode of *The Great War* or *The World at War*, to ask oneself to what extent the actuality footage really earns its place in the programme. So much of *The Great War* seemed to consist of soldiers marching along roads, with the burden of the argument being carried in the narration or by a characteristic (in the bourgeois historiographical tradition) recourse to non-filmic 'documents' — orders of the day, extracts from memoirs, personal testimony, newspaper reports, and so on. At the same time, that same actuality footage had to be subjected to a complex technical process (the double printing of every third frame) *in order that it function ideologically as actuality film* (i.e., look to a modern audience like a recording of 'real' life and not evoke the somewhat comic response which usually greets the jerky movements of figures in films cranked at (usually) sixteen frames per second).

The authority of the *Real* is asserted also in reconstruction footage, that is those sequences (in tele-history programmes dealing with pre-twentieth century periods) where dramatic reconstruction rather than contemporary pictorial artefacts are used. The overwhelming impression of such sequences is their *disproportionate length* in relation to the historical points being made (e.g., the cloth-dyeing sequence in the 'In Darkest Africa' episode of *The British Empire* and the tribal dancing sequence in the 'Free Paper Come' episode of *The Fight Against Slavery*) and their *autonomy*, that is their tendency to function in relation to criteria internal to themselves rather than in relationship with the

overall historical argument which the programme is making. In the 'Free Paper Come' episode of *The Fight Against Slavery* there is a dramatic reconstruction of the 1832 uprising of slaves in Jamaica and of the reprisals which followed its being put down. One sequence shows the burning of a slave village (doubtless a historical fact) which could have been signified in several ways. What is interesting is the length of the sequence and, particularly, its autonomy: the burning huts are shown, by means of dissolves, at various stages of the process of burning, with the final dissolve being to a heap of charred embers with appropriate noises on the soundtrack. The strongest impression is conveyed of the makers of that particular sequence operating within certain autonomous conceptions of drama, rather than within a conception of a form within which history is signified; i.e., they are more interested in arresting visual effects than in historical knowledge.

The manifestation of the phenomenon we have described as the authority of the *Real* with which readers may be most familiar is that which places the narrator of the series in the actual location of the events he is narrating. Thus Dr Bronowski, in the space of one programme, may be seen on the Icelandic coast, in his California home, on Easter Island and in the Caves of Altamira; Magnus Magnusson may move from Petra by way of Herodium and Qumran to Masada; and John Kenneth Galbraith be spirited from Edinburgh via Paris to Quebec. This locating of the narrator in the actual substance of his narration offers a quasi-talismanic guarantee of truth: the *place* actually exists, therefore what is said must be true. At the same time there is in this phenomenon a sub-text which has nothing to do with *history* but a lot to do with *television*. What we are seeing is television—and specifically the well-financed co-production series—displaying its *resources*. *

It is a running argument of this monograph that the categories of television and those of bourgeois historiography—while relatively autonomous within their own practices—will (since these practices constitute regional activities of the superstructure as a whole) be related at the philosophical level and will therefore reinforce each other. The impulse towards visible and tangible evidence in bourgeois historiography, expressed in tele-history as a fetishisation of actuality film, is re-echoed in several of the categories described by Stuart Hall (see chapter 3) as contituting 'good television', particularly the demand that 'good television' capture 'the ongoing form of real life'.

Notes

1. 'The Historian and His Facts' in *What Is History?*, Penguin, Harmondsworth 1961, p.9.
2. See, for example, Goldstein, op.cit. pp.63-91.

*I must confess to experiencing certain irreverent feelings when watching series such as *Civilisation, The Ascent of Man* and *The Age of Uncertainty* which emerged as a recurrent disbelieving cry: 'Christ, did they fly Clark/Bronowski/Galbraith and a film crew all the way there just to say *that*?'.

7 The Idea of Progress and the Inviolability of Chronological Monism

All this talk about the decline of civilisation means only that university professors used to have domestic servants and now do their own washing up. (A.J.P. Taylor, as quoted by E.H. Carr)

The idea of *Progress*, of mankind or human society developing from lower to higher forms,[1] has had a long, convoluted and complex history. As a recent historian has written:

> The history of the belief in general progress from the early eighteenth century down to the 1880s is, on the whole, a history of successively broader and more far-ranging conceptions of the progress of mankind . . .
>
> The one ingredient most lacking in eighteenth century French ideas of progress — an organic sense of history as a process of continuous development — was incorporated into progressivist theory late in the eighteenth century and throughout the nineteenth century by thinkers from every national tradition. Three principal strategies may be distinguished, each working to the same broad ends: the approach of German historical idealism, which blended a radically immanentised version of the Christian doctrine of divine providence with an ingenious amalgam of rationalist and romantic thought; the approach of the new sociology, which took as its project the discovery of the scientific laws of progress; and the approach of the evolutionists, which accepted the positivism of sociology, but sought to root the idea of human progress in a larger scheme of biological or cosmic progress.[2]

It is hardly to be expected that a society which has lived through two world wars, through the insights of psychoanalysis and through the fashionable despair and cynicism of certain twentieth century 'philosophies' would write history with quite the same degree of positivist optimism which characterised bourgeois historiography in the second half of the nineteenth century. Nevertheless, the *humanist* progressivism of the late nineteenth century is still apparent in tele-history and in television historical drama. It is apparent in the thrust of a series such as *The Fight Against Slavery*,[3] in the title of *The Ascent of Man* and, indeed, in that programme's overall philosophical stance.

Dr Bronowski's classical progressivism is evident from the following excerpts from his narration of 'Lower than the Angels', the opening programme in *The Ascent of Man* [4] series:

> That series of inventions by which Man, from age to age, has remade

his environment, is a different kind of evolution, not biological but cultural evolution. I call that brilliant series of cultural peaks The Ascent of Man . . .

and

. . . yet human achievement, and science in particular, is not a museum of finished constructions; it's a progress in which the first experiments of the alchemists also have a formative place . . .

Also the Euro-centrism and, in the context of British television, the Anglo-centrism of the progressivist tradition is apparent both in tele-history (e.g., the BBC's *The British Empire, Churchill's People, The Great War*) and in television historical drama such as *Edward the Seventh* and *The Explorers* (BBC).*

It is important to distinguish between a conception of historical progress which is positivist, humanistic and Euro-centric—the conception informing British tele-history and television historical drama—and one which is none of these. The latter has been outlined by E.H. Carr:

. . . no sane person ever believed in a kind of progress which advanced in an unbroken straight line without reverses and deviations and breaks in continuity, so that even the sharpest reverse is not necessarily fatal to the belief. Clearly there are periods of regression as well as periods of progress. Moreover, it would be rash to assume that, after a retreat, the advance will be resumed from the same point or along the same line. Hegel's or Marx's four or three civilisations, Toynbee's twenty-one civilisations, the theory of a life-cycle of civilisations passing through rise, decline and fall—such schemes make no sense in themselves. But they are symptomatic of the observed fact that the effort which is needed to drive civilisation forward dies away in one place and is later resumed at another, so that whatever progress we can observe in history is certainly not continuous either in time or in place. Indeed, if I were addicted to formulating laws of history, one such law would be to the effect that the group—call it a class, a nation, a continent, a civilisation, what you will—which plays the leading role in the advance of civilisation in one period is unlikely to play a similar role in the next period, and this for the good reason that it will be too deeply imbued with the traditions, interests, and ideologies of the earlier period to be able to adapt itself to the demands and conditions of the next period . . .[5]

Clearly, the conception of progress outlined by Carr is much closer to the Marxian sense of progress than to the bourgeois liberal sense, although both senses emerged out of the same nineteenth century ethos. However, a conception of progress based on a scientific theory of the succession of modes

*With regard to the internal history of Britain and the British perspective on history generally, the point of view is almost invariably that of middle-class Englishmen. The perspectives of the non-middle class and Celtic peoples of Britain (not to mention those of more recently constituted minority groups in the UK) are all but ignored except on 'regional' programmes.

of production is very different from one based on notions of the perfectibility of individual men and women and a *progressivist* view of history such as the former is not the same as a *teleological* view of history—i.e., the view that history moves to a pre-ordained end.

The Marxian conception of progress is not conspicuously visible in British tele-history and television historical drama.

Closely associated with the idea of progress, in the positivist and humanist sense, is that of historical time as being unproblematic, linear, evenly developmental. Insofar as this has been theorised, it is perhaps best glimpsed in Hegel's conception of history in which time is *continuous* and *homogeneous*—thereby posing for the historian the problem of *periodising* accurately when one dialectical totality ends and another begins—and in which time is *contemporaneous*—thereby allowing the historian to take an 'essential section' and be certain that in so doing he can consider all the elements of the dialectical totality which, in the Hegelian conception, co-exist at one and the same time.

This view of historical time (or at any rate vulgarised versions of it)—by virtue of its following very closely the contours of lived experience—suffuses tele-history and television historical drama. However, it is a view which has been challenged both by Marxist and non-Marxist historians and philosophers, either in their own practice as historians or in their reflections about history-writing. In many respects it is one of the most complex, difficult and controversial areas of concern within contemporary Marxism,[6] but for the purposes of this monograph a simplified view of both the theory and practice of this counter position can be given to demonstrate its absence from tele-history and television historical drama and to begin to pose possible alternatives to the dominant practices in these areas.

At the theoretical level, Althusser writes:

> As a first approximation, we can argue from the specific structure of the Marxist whole that it is no longer possible to think the process of the development of the different levels of the whole *in the same historical time*. Each of these different 'levels' does not have the same type of historical existence. On the contrary, we have to assign to each level a *peculiar time*, relatively autonomous and hence relatively independent, even in its dependence, of the 'times' of the other levels. We can and must say: for each mode of production there is a peculiar time and history, punctuated in a specific way by the development of the productive forces; the relations of production have their peculiar time and history, punctuated in a specific way; the political superstructure has its own history . . .; philosophy has its own time and history; aesthetic productions have their own time and history . . .; scientific formations have their own time and history, etc.[7]

As well as pointing to some of the problems of this formulation, Vilar demonstrates what the application of this notion of *differential temporality* might look like in concrete history writing:

If one interprets Labrousse* as saying: the French Revolution was born of a 'fusion' between a *long time*—the economic expansion of the eighteenth century—a *medium* time—the intercycle of depression, 1774-88—and a *short time*—the price crises of 1789 which culminated (almost too perfectly) in the *seasonal* paroxysm of July 1789, then it looks as if the demonstration is a mechanistic explanation of the revolution which shuffles together linear times as if they amounted to a casual concatenation. But is this what he says?

In fact, the statistically observable *short cycle* which pulsates in the economic *and social* reality of the French eighteenth century is *the original cycle of the feudal mode of production*, in which: (i) the basis of production remains agricultural; (ii) the basic productive techniques do not yet dominate the stochastic cycle of production; (iii) the dues levied on the producers *should* vary according to the amount produced; (iv) charity and taxation *should* cushion the worst forms of misery, in a bad year.

However, this pre-capitalist tempo already co-exists with others in the eighteenth century which though not yet typical of the future mode of production (like the 'industrial cycle' for instance) pave the way for it and are part of it: (i) a *long period* of preparatory accumulation of money capital, directly or indirectly *colonial* in origin, which creates a moneyed bourgeoisie and 'bourgeoisifies' part of the nobility; (ii) the *medium-term* possibility of *commercial depressions* (market crises, price depressions) affecting and upsetting growing numbers of farmers, proprietors and entrepreneurs whose products have entered the commercial circuit and become 'commodities'—so many social strata interested in legal equality, free markets, and the end of feudal structures; (iii) lastly, the aggravation of 'old-style crises' *in the short run*, since though they are less lethal than in the days of famines, the new speculation on shortages which they provoke is less restrained by administrative taxes and ecclesiastical redistributions, and they therefore pauperise and proletarianise the masses more than ever and turn the poorer peasants against both feudal or royal levies and market freedom.

What better example than this could one find of an 'interlacing of times' as the 'process of development of a mode of production' or even as a transition from one mode to another—this convergence of 'specific temporalities' which in July-August 1789 resulted in the famous 'event' that overthrew the whole juridical and political structure of society?[8]

I know of no programme, either of tele-history or television historical drama, which incorporates into its structure an awareness of the notion of differential temporality or even signals the existence of a problem in that regard. Rather the problem is repressed and displaced onto other terrain.

A good example of an area where the notion of differential temporality might have usefully come into play is in the episode of *Edward the Seventh* entitled 'The Royal Quadrille' (discussed in detail in chapter 8). In this

---

*An important French economic historian.

33

programme, several crowned heads of Europe are gathered together, but most notably the monarchs of Britain, Germany and Russia. As presented in the programme their forms of monarchy 'bathe in a more or less continuous and homogeneous temporality'. To be sure, the constitutional monarchy of Britain and the (in themselves quite distinct) absolute monarchies of Germany and Russia existed contemporaneously. What is wholly repressed in the programme, however, is that these diverse forms of monarchy took their highly individual character from the fact that they constituted different stages of development in relation to the transition from feudal to bourgeois society within their own concrete social formations.[9] Characteristically in this bourgeois television drama, these structural differences between British constitutionalism and the Russian and German absolutisms are repressed and displaced to reappear in the form of personality differences among the monarchs and their circles, with the added suggestion that the constitutional reasonableness of Edward VII is a reflection of the innate decency and good sense of the British people.

Analogously, the notion of *differential temporality* might have thrown light—in *The Fight Against Slavery*—upon the question of why it took until 1834 to get slavery abolished in the British West Indies. The key element, of course, is the different times of the British economic formation and the British political formation as expressed in the Westminster parliament. The Industrial Revolution decisively reduced British economic dependence on slavery sometime after 1780, but the different tempo of the British polity did not render the West Indian interest insignificant in the Westminster parliament until 1832. Such notions might have been nearer the mark—in *The Fight Against Slavery*—than, as was foregrounded, the moral earnestness of Wilberforce, Clarkson and other abolitionists and the highly rhetorical debates in the Westminster Parliament.

The argument of the monograph to this point has sought to adumbrate the main categories which inform tele-history programmes and television historical drama and demonstrate empirically the presence of these categories and the corresponding absence of categories from theoretical positions other than the bourgeois, in particular the historical materialist position.

There is a danger that the argument might be interpreted as suggesting that phenomena such as the centrality of the individual, the authority of the 'real' and the universality of chronological monism are philosophical categories informing the programmes but which lie so deep and inert as to have no real ideological force. This would be a misinterpretation of the argument which is that these categories provide the absolutely necessary philosophical sub-soil for the growth of intensely active ideological interventions into the here and now.

This is best demonstrated by detailed consideration in the next two chapters of two pieces of historical drama, the one directly foregrounding historical personages and events, the other operating primarily as 'fiction' but refracting particular historical personages and events through its ongoing, serial structure.

Notes

1. Impulses towards this view, if not the term itself, have been detected in classical and medieval Christian writings with an immense thrust towards the theorising of the position coming with the Renaissance and the Scientific Revolution. See, for example, R. A. Tsanoff, *Civilisation and Progress*, University of Kentucky, 1971.
2. W. Warren Wagar, *Good Tidings: the Belief in Progress from Darwin to Marcuse*, London, 1972, pp. 18-19.
3. In this context, however, there is something of a tension between the *mise en scène* of the series as realised on the screen—e.g., the foregrounding of successive parliamentary debates leading to the Abolition Bill and the autonomy of the dramatic reconstructions—and the writing and commentary of Evan Jones, himself a black Jamaican, with the final image of the series being a hangman's noose and an ex-slave's voice-over saying 'Slavery don't finish yet.'.
4. It is a point of view which is best argued in a different context from this monograph, but in many respects Bronowski's is the very archetype of the bourgeois sensibility, particularly in its anthropocentrism and idealism and its appropriation of high art (Leonardo, Shakespeare) as mystical evidence of 'the ascent of man'. This impulse towards the appropriation of high art and the correspondingly high rhetoric of Bronowski's gestures and argument put a severe strain on the pictorial form of *The Ascent of Man*, impelling it to baroque and 'elemental' imagery (sea, sun, rock) and causing it to locate itself within the most inflated popular pictorial tradition available, that of the cinematic biblical epic (cf. the opening credits and the theme music of the series).
5. E. H. Carr, 'History as Progress' in *What is History?*, Penguin Books, London, 1971 ed., p. 116.
6. See, for instance, the philosopher Louis Althusser's 'The Errors of Classical Economics' in *Reading Capital*, New Left Books, London 1970, and the historian Pierre Vilar's answer to this, 'Marxist History, a History in the Making' in *New Left Review*, No. 80, July/August 1973.
7. Louis Althusser, 'The Errors of Classical Economics', op. cit., pp. 99-100.
8. Pierre Vilar, 'Marxist History, a History in the Making', op. cit., pp. 88-89.
9. The *complexity* of this issue cannot be overstressed. It is not simply a question of Britain being more 'advanced' than Germany and Russia, but of the peculiar mixes of each social formulation. Thus, in the case of Britain, it is not its being 'advanced' which allowed the trappings of constitutionalism to be displayed, but in a very special sense its being *retarded*, i.e., its retention of certain features of feudalism. For a fully developed version of this argument see Tom Nairn, 'The Twilight of the British State' in *New Left Review*, Nos. 101-102, February/April 1977.

8 Constructing the Past for the Ideological Needs of the Present

. . . however remote in time events thus recounted may seem to be, the history, in reality, refers to present needs and present situations wherein those events vibrate. (Benedetto Croce)

All television drama is ideological. Take a random episode of a random series—the episode of *The Fantastic Journey* entitled 'A Dream of Conquest'. The band of time-travellers, searching for their own 'dimension', enter another one to find a group of heavily uniformed soldiers tormenting a strange animal, half-dog, half-bear. They too are arrested and taken to the soldiers' leader, Tarant, whose discourse consists largely of rhetorical dreams of military conquest. He is treacherous and ruthless and the key words in his vocabulary are 'power', 'discipline' and even 'the new order'. It is readily apparent that—although the episode is set in the future—the main point of reference is the recent past, specifically the accoutrements and rhetoric of Fascism, particularly of the German type. The ensemble of visual and verbal images whereby Fascism has been rendered in popular art functions as a set of tokens for exchange between programme makers and audiences, the tokens being redeemed by appropriate emotional responses to particular characters and situations.

The Fantastic Journey's ideological project is somewhat blunt and ill-defined, pro-'democracy', anti-fascist. The ideological projects of the kind of historical drama under scrutiny in this monograph are—while operating to the same ideological ends and drawing upon the same ensemble of tokens—altogether sharper and more precise.

The central ideological project* of *Edward the Seventh*, for example, is no less than the humanisation of the British monarchy. It is hardly accidental that *Edward the Seventh* is contemporaneous with other mechanisms designed to further the same ideological ends, e.g., increasingly informal official photographs of the monarch and the royal family, increased public access—by way of the media—to previously closed areas of royal activity, royal 'walkabouts' on foreign and domestic tours.

It is useful to consider a single episode[1] of *Edward the Seventh* in some detail, for as well as following through the quite complex workings of the central ideological project, it is possible to discern in operation a secondary ideological project of great force and many of the mechanisms and concepts, discussed elsewhere in this monograph, which together constitute bourgeois superstructural activity.

*The term 'project' refers to the thrust of the programme, the way it presents itself to be read, and **not** to the conscious intentions of the programme makers.

The episode entitled 'The Royal Quadrille' is exemplary in the free passage it allows bourgeois ideology. The almost inevitable pre-credit sequence presents three gentlemen out walking on a flat expanse of land. The audience learns that they are in fact Edward, the Prince of Wales, known as 'Bertie'; Alexander III, Czar of Russia, who will be addressed as 'Sacha'; and George III of the Hellenes. Part of the *frisson* provided by the series, and an important aspect of its central ideological project, is to hear the crowned heads of Europe and their wives address each other by their first names and nicknames.

Sacha and George are arguing as to the direction their destination lies in. Bertie remains aloof from the argument and when a passing carter informs them that it can be reached by going in both directions, Bertie intervenes to assure his arguing relatives that they are both right. Thus the important ideological point is made that the heir to the British throne is a mediator. Given a lift in the cart, the three introduce themselves to the disbelieving carter who replies, 'And I'm Napoleon Bonaparte'. The humanisation of monarchy is under way.

The royals are holidaying at the summer palace of the Danish king and there follows a scene of remarkable ideological interest since it continues the implementation of the central ideological project, the humanisation of the British monarchy, and signals the beginning of the secondary ideological project—the subsuming of Russia and Germany within the single category of *totalitarianism*.

The Czarina—the year is 1887—talks of being 'a prisoner in my own house' and of 'security men' being everywhere. Now professional historians are aware that one of the cardinal sins of historiography is to impute to figures in one historical epoch the feelings and views of the writer's own time. While conceding the historical presence of a secret police in Russia for over two centuries before the Revolution, it cannot be argued but that the term 'security men' is anachronistic, carrying with it connotations of the Stalin era and having the precise ideological function of constructing Russia as a land of timeless tyranny which transcends any particular political formation. As will be demonstrated, this will connect with the programme's later construction of Germany, but the scene in which the phrase is spoken has the primary ideological function of opposing 'totalitarian' Russia to 'democratic' Britain. As Sacha raves on about his father's assassination and the necessity of security, Bertie genially interposes, 'Don't you think there could be greater security in the love of your people than in the strength of your police force?' Such is the strength of bourgeois ideology that it can show its hand openly! Sacha does not take too kindly to this intervention and goes off in a huff, whereupon George of the Hellenes offers some explanation of his behaviour and at the same time advances the secondary ideological project of the programme— 'He's afraid of Bismarck and German expansion', to which one of the ladies replies, 'Bismarck can't last forever and when he dies Vicky and Fritz will stop German expansionism . . .', 'Vicky' and 'Fritz' being Victoria and Frederick, briefly to be the Empress and Kaiser of Germany.

The construction of Germany as imperialistic and totalitarian proceeds with the entry of Wilhelm, grandson of the current Kaiser and himself to be the Kaiser at the time of Germany's entry to the First World War. He is played by an actor of decidedly 'Aryan' type and is presented as arrogant and unbending with much heel-clicking. He greets Bertie with the words 'Mr Gladstone has embarked on a holy war and sent a military force to Alexandria . . .' (Throughout this scene and the programme—and, indeed, the series—the bourgeois historiographical category of the *individual* holds sway with great clarity, as is evident by the remarks quoted above.)

The humanisation of the British monarchy continues by the signalling—through his adulterous liaison with, in this episode, Lily Langtry—that Bertie is a man with all the normal human appetites and by the signalling that his human virtues are constrained by the demands made on him as heir to the throne by the Constitution. His *courage* is signalled in the scene in which he argues with his mother, Queen Victoria*, when she forbids him joining the military expedition to Alexandria and his *concern for the poor* is signalled in the, quite lengthy, scenes dealing with his involvement with the Royal Commission on the Housing of the Working Classes and in his stated wish to vote in the House of Lords for a reform bill in that connection. However, having achieved the ideological point of the programme by the very act of displaying this concern, he bows to the ideology that 'the Royal family must be above politics'.

The secondary project of the programme, relating to the construction of Germany as imperialistic and totalitarian, is achieved through the traducing of Bismarck and the reiteration of his baleful influence on the young 'Willie' (Wilhelm). In a dramatically quite skilful construction, Bismarck is seen at a ball with the various royal families of Europe. He stands stern and aloof, a mixture of Machiavelli and Svengali, literally like some monstrous puppet-master, watching the dancers and becoming perturbed when he sees dancing together a prince and princess whose continued liaison 'will throw Russia into the arms of France'. Thus, in television historical drama, are the complexities of European political alliances signalled.

Just as the pre-credit sequence of 'The Royal Quadrille' was heavy with ideological import, so too is the final scene. Following the death of Kaiser Frederick III, his British-born widow, Victoria, laments the influence of Bismarck on her son, now Kaiser Wilhelm. She talks of being afraid of her own son and, in the very same words as, earlier, the Czarina spoke of her own situation in Russia, of being 'virtually a prisoner in my own house'. With these

---

*There is another dimension of television historical drama which—though not *directly* related to the categories of bourgeois historiography—is nevertheless ultimately connected with the philosophical category of the *individual*. This dimension is the formal centrality of the 'great performance'. Television historical drama is rich in these, which consist of the kind of highly baroque performances which are the very reverse of naturalistic and which seem to gain their popular acclaim from audience awareness of the fact that the player is visibly 'performing'. Such performances regularly win television drama awards and are typified by Derek Jacobi's Claudius, and in the series under discussion here, Annette Crosbie's Queen Victoria.

This is a good example of the categories of bourgeois historiography and bourgeois (television) drama reinforcing each other.

words, the secondary project subsuming Russia and Germany within the same category of *totalitarianism* is achieved. It is a project which, of course, operates dialectically with the central ideological project of the programme, the humanisation of the British monarchy.

It remains only to consolidate that secondary ideological project by the kind of dire foreboding which is a feature of bourgeois historical drama. The young Kaiser Wilhelm enters, coldly waits for the bows and curtsies of his international relatives, including his mother, calls her to his arm and exits with music rising on the soundtrack. The music bears a close family resemblance to *Deutschland Uber Alles.*

Notes

1. Interestingly, the same ideological impulses and projects are evident in other episodes of the series where key personnel (e.g. the writer) are different from those working on the episode discussed in detail here. The episode entitled 'The Peacemaker' shows Edward as individually responsible for putting together the Triple Entente with France and Russia, in the teeth of the reluctance of the British government and diplomatic service, and for initiating reform of the army and navy. The continuity of the ideological projects discussed in this chapter is evident from the following remarks to, about or by Edward in 'The Peacemaker': a French actress—'we are only a republic because we do not have royalty like you'; an admiring aide—'he doesn't have the kind of mind to read complicated reports, but he *can* read people'; Edward, *vis-à-vis* Churchill and Lloyd George — 'it is easy to be radical: what this country needs is men of moderation and commonsense'; and Edward to the Czarina of Russia—'letting your people have a share in the government of their country is not "giving in" . . .'.

Typically, the diplomatic manoeuvres leading up to the Triple Entente are presented (within the ideology of the individual as maker of history) as the 'moderate' Edward organising peace in the face of the 'extremist' Wilhelm, rather than as the various European nation states getting ready for a showdown over their diverse (intra- and extra-European) imperialisms.

9 The Cannibalisation of History

Elderly people in our culture are frequently oriented towards the past, the time of their vigour and power, and resist the future as a threat. It is probable that a whole culture in an advanced state of loss of relative power and disintegration may thus have a dominant orientation towards a lost golden age while life is lived sluggishly along in the present. (R. S. Lynd)

Edward the Seventh represents a more direct form of quite a common phenomenon in television—the recurrent return to the late Victorian, Edwardian and Georgian periods as milieux for drama. This is evident in, among others, *The Forsyte Saga* (BBC), *The Duchess of Duke Street* (BBC), *Raffles* (BBC), *Clayhanger* (Granada) and the Wodehouse and Chesterton adaptations, although important distinctions have to be made between programmes which primarily signify 'nostalgia for an earlier period' and those which signify 'adaptation of a Great Literary Classic'.

In many respects the archetype of the former is *Upstairs, Downstairs* (LWT), certainly one of the most successful, in commercial and critical terms, both in the UK and abroad (for example, it won an 'Emmy'—the US television drama award). Clearly, the series' excellence in terms of the norms of bourgeois drama—the writing, the playing, the production values—are important elements in its success, but the argument of this monograph is that all television (including drama) fulfils an ideological function and that there will be a relationship between the popularity of a programme and the extent to which it reinforces the ideological position of the majority audience.

To be speculative, it seems reasonable to suppose that a society going through a period of historical transition and finding it immensely painful and disorienting will therefore tend to recreate, in some at least of its art, images of more (apparently) settled times, especially times in which the self-image of the society as a whole was buoyant and optimistic. For post-war Britain, faced as it is with adjustment to being a post-colonial power, a mediocre economic performer, a multi-racial society and a society in which the consensus of acceptable social and political behaviour is fragmenting (all, of course, factors which are intimately inter-related), what better ideological choice, in its art, than to return to the period of the zenith of bourgeois and imperial power or to immediately succeeding periods in which the façade of that power appeared convincing.

As the quotation from Croce (which provides the rubric for the previous chapter) indicates, no matter what period history-writing or historical drama is ostensibly dealing with, in reality it is providing for the ideological needs of the present. Thus, one of the projects of programmes such as *Upstairs, Downstairs* is the feeding of a dangerous contemporary nostalgia for more settled times but it is doubtful if a project of such generality would sustain the series over the lengthy period of its run. One would expect to find, therefore, that—just like *Edward the Seventh*—*Upstairs, Downstairs* will offer more concrete ideological guidance to the 'problems' of today. Unlike *Edward the Seventh*, it does not deal regularly with historical personages (although this or that historical figure will appear in a particular episode). It operates ideologically, therefore, by what could be called a process of cannibalising history, by taking particular historical events and offering ideological guidance by refracting them through the on-going, well-signified, and well-understood value-system of the series.

This value-system is well-known. The series deals with a particular household through the late Victorian, Edwardian and early Georgian periods. With an unerring ideological accuracy which presents the curious mixture of the aristocratic and the *haute bourgeois* lying at the heart of the British system of class power, the series has the (original) wife coming from an aristocratic family and the husband (the absolutely key ideological force in the series) as an upper middle-class Tory Member of Parliament. The mix of the aristocratic and the bourgeois is skilfully retained, after the death of the first wife on the Titanic, by ennobling the husband and having him marry a bourgeois woman.

It is the *household* which the series deals with. We come to know the husband and wife, their children, relations and friends *and* the domestic staff—as hierarchical in their way as the family above stairs—with the butler and cook functioning as the analogues of the husband and wife. (This analogue was made explicit at the end of the series by the marriage of the butler and the cook.)

We see the family and the servants in their own lives and in their inter-relationships and despite the sometimes quite extreme dramatic situations—bereavement, suicide, unwanted pregnancy, etc.—and despite the fact that both classes are seen 'warts and all'—the overall mood of the series is one of celebration of the relatively cosy stability of a society in which everyone (certainly the regular 'characters') knows his/her place, accepts it and is treated with 'dignity' and 'kindness' within it.*

Within the range of 'characters' the audience has come to know (and, dare one say it, 'love') expectations are created as to what their responses will be to any situation or event. Thus, the aristocratic wife, though kind and charming, is a little 'old-fashioned'; the somewhat feckless son is liable to opt for apocalyptic solutions to problems (he eventually blows his brains out); the

---

*It is ideologically interesting that such a point of view can still be given currency without too many raised eyebrows. The recent highly popular American series *Roots* gains much of its *frisson* from its presentation of the whites operating a slave society as though it were a natural state of affairs and their consequent blindness to even the most basic human aspirations of the blacks.

Scottish butler effectively conceals his basic humanity under a sharp sense of class proprieties further stiffened with Calvinism; and the spinsterish lady's maid, though loyal and plucky, does not pretend to understand the complexities of society. Standing like a rock at the centre of the series—the ideological rock, that is, not the dramatic rock, a role which oscilllates between the Scottish butler and the lady's maid—is the father of the house. It is invariably he who finds the 'sensible' solution to crises where solutions are possible: it is he who is our ideological guide through history and, by extension, through the problems of the present.

This, then, is the ongoing framework through which history is mediated. It is useful, therefore, to consider in some detail how a particular historical event (the General Strike) is refracted through *Upstairs, Downstairs.*

The title of the episode within which the General Strike figures—'The Nine Day Wonder'—is in itself ideologically interesting and gives some indication of where the programme's sympathies will eventually lie. The question is broached very early on in the episode between Hudson (the Scottish butler) and James (the feckless son of the house). Their attitudes would be recognised as characteristic by the audience:

> James: Coal fires in May. Miners out on strike and the rest of the country all jumping on the bandwagon. What a mess!
> Hudson: I gather there's still hope of a settlement, sir. The trade union leaders are at Downing Street at this very moment. It's just been announced on the wireless.
> James: But it shouldn't have been allowed to come to this. A General Strike is a direct affront to the government. An insult to democracy. It should be forbidden. By law.
> Hudson: I agree with you, sir. I feel deeply ashamed of my fellow working man.
> James: You, Hudson? You've got nothing to be ashamed about. Not your fault. It's men like Cook and Thomas and Bevin . . . so called leaders . . . having the nerve to hold the country to ransom . . . and threaten the liberty of ordinary, decent people . . .

The phrase used by James 'hold the country to ransom' was a key media phrase during the miners' strike of 1973/74 which led to the fall of the then Conservative government. The use of this phrase signals to us that the central ideological project of the programme has to do with attitudes not to the General Strike *per se*, but to working class militancy in our society here and now.

The other figures in the household talk and act in character: Frederick, an ambitious footman and wartime batman to James, talks with relish about the armed forces getting ready for the fight to come (he and James 'scab' on the buses); Hudson, claiming Winston Churchill as authority, asserts that 'all miners are reds' to which Ruby, the kitchen-maid, retorts that her uncle is a miner and he's not a red. Lady Prudence, a close family friend, having remarked on the lack of consideration of the strikers, throws her home open to Oxbridge students 'scabbing' on the trains; and Georgina (the father's ward), while censorious of James' pomposity, nevertheless organises deliveries of the British Gazette.

Against such blinkered responses, Richard (the father) enters and begins to fulfil his ongoing ideological role of mediator and compromiser, of incarnating Social Democracy. Asked what had happened to the last-minute talks between the TUC leaders and the Government, Richard replies:

> Ended in confusion. Apparently some Daily Mail printers refused to print an article which condemned the strike. Baldwin got to hear about it and sent Thomas and company home. Far too hasty in my opinion. Nobody wants this wretched strike. They were looking to Baldwin to help them save face.

Significantly, Richard's first condemnation is not of the strikers, but of the government and when James talks of the masses as about to do a Russia 1917 job, he replies:

> I don't go as far as James, but there's certainly a strong feeling of solidarity in the working classes, rather like the early days of the war.

This construction by Richard of the British working-class as decent, human and patriotic will be taken up and developed when Ruby's uncle, already referred to, comes to the house with a fellow-miner, both of them having come from the north of England on a miners' delegation. However, at the same time as Richard is fulfilling his *ideological* role of mediator and spokesman for Social Democracy against blind reaction, he nevertheless fulfils his *class* role of complimenting everyone, family and servants alike, on the measures they are taking to keep things going—'we must all pull together, do what we can'— and his last act before leaving for the House ('to see what I can contribute') is to give Hudson permission to enlist as a special constable. The point at which Richard's *class* role and his *ideological* role fuse—or, more accurately, where the contradictions lie uneasily together — is when he addresses the servants collectively about the strike and sounds most like a social democratic politician:

> Now I don't want to go into all the issues . . . who's right and who's wrong in this dispute. I'm sure you realise very clearly that for the future life and prosperity of this country the strike must not be allowed to succeed. Nor, on the other hand, must it develop into a violent and bitter struggle between the classes. A solution will be found. In the meantime, we must all show restraint, patience, good humour . . .

The scene in which this speech occurs is reminiscent of those in English war films when the commanding officer gives a morale-boosting talk to the 'other ranks'. Richard even compliments Ruby (standing in as cook for the absent Mrs Bridges) on the quality of last night's apple dumplings. It is a mark of the ongoing power of the series—of the 'human' capital accumulated by the characters in previous episodes—that this scene can be played straight with no hint of the programme-makers being distantiated from the event they are portraying.

All episodes of *Upstairs, Downstairs* function centrally in terms of personal crises or sharp inter-personal animosities. The General Strike acts as a catalyst for the tensions adumbrated in the series as a whole. Thus James and Georgina clash over his blood-curdling views on the outcome of the strike and the class-collaborative figures of Hudson and Frederick clash with others such as Edward the chauffeur and Ruby the kitchen-maid (significantly, the less intelligent characters as presented in the series) who have a sharper sense of class solidarity.

Despite the overall project of nostalgically presenting a society which is relatively cosy and stable, *Upstairs, Downstairs* does have a 'then' and 'now' view of class. It is part of a wider feature of bourgeois historical drama as a whole (see also the discussion of *Edward the Seventh* in chapter 8) that the past is re-read from the point of view and with all the knowledge of the present. This is in evidence when Frederick, having had explained to him why working men want other working men like himself to support the strike, replies 'That's daft, that is. I mean . . . we got no grievances . . . have we?' and, more obviously, when Edward begins to get uneasy about his non-participation:

> . . . you see all these people standin' round in groups, bus drivers, engine drivers and that, all out of work, stickin' up for the miners . . . well it's started me thinkin' maybe we should be out there with 'em . . .

However, he is brought into line by Rose's angry appeal to his sense of loyalty to His Lordship, the evidence of 'clever people . . . on the wireless . . . sayin' it's wicked and causin' misery', and her invocation of Mr Hudson's name.

The programme's presentation of Ruby's Uncle Len and his friend Arnold is absolutely central in terms of its ideological project of valorising Social Democracy. The two are introduced at a key dramatic moment (i.e., just prior to the commercial break) and their presentation connects with the working out of the same project through the figure of Richard:

> Arnold: We're not fightin' constitution. We're fighting for bread. We're makin' no demands. We're not chasin' moon.
> Len: Just a simple livin' wage.
> Rose: But I thought it was the miners started it.
> Arnold: Nay . . . it were forced on us . . . when owners locked us out and government supported them.
> Len: But church says we're right. Archbishop of Canterbury himself . . .
> Arnold: Aye . . . and BBC won't broadcast what he says. Government won't let 'im. Call ourselves a Christian country!

The programme presents the miners (in terms of the writing, casting and playing) as human, dignified, a-political and, by implication, Christian. They explicitly deny that they are communists and when a row follows the appearance of Hudson on the scene, the audience is left with the distinct feeling that the miners emerge with greater credit and dignity than Hudson. When they return to the house towards the end of the programme and learn that the strike has been called off they leave vowing not to accept the decision, thereby signalling the actuality of the sequel, the miners' six-month struggle at the end of which starvation drove them back to work.

Richard's interventions during the rest of the programme show him working for the compromise which will 'preserve the dignity' of all concerned (and preserve the social status quo), his eulogies on strong trade unions and civil liberties and, finally, his epilogue on the strike:

> It was a fair trial of strength. Both sides kept their heads. There was loyalty, self-sacrifice, very little anger. I think the whole nation can be proud . . .*

It is interesting that a programme such as *Upstairs, Downstairs,* using the mechanisms which it has carefully constructed since its inception, should refract history so as to land on exactly the same ideological spot as a series, *Edward the Seventh,* which is more directly concerned with history. However, to revert to the ongoing argument of this monograph, they both represent superstructural activity in contemporary Britain and, as such, will necessarily show ideological similarities. The humanisation of the British monarchy and the valorisation of Social Democracy constitute the best available terrain on which to fight for the maintenance of the socio-economic status quo.

I have tried to demonstrate the process of ideological intervention in the here and now primarily by discussion of historical drama since this (like all the other 'entertainment' areas of television) is the area where we are most off our guard. The case could, of course, have been argued with equal detail in relation to tele-history programmes. I am aware of no tele-history programme in which a Marxist historian has been invited to make a substantial contribution and the inflated, internationally co-produced 'personal' histories have all been by figures, conservative like Clark or reformist like Galbraith, who rest easily within the parameters of Social Democracy. That the great bulk of tele-history hits the same ideological mark as *Edward the Seventh* and *Upstairs, Downstairs* may be demonstrated by looking closely at a short sequence from the 'Home From Home' episode of Alistair Cooke's *America.*

Having dealt with the seventeenth century English settlements in Virginia and praised their contribution to the emergence of American democracy, Cooke goes north to New England to examine the English Puritan settlements there. In a sequence which is remarkable for the display of many of the bourgeois tele-historical mechanisms discussed in this monograph (the fetishisation of the individual and of historical location, the aspiration of tele-history to drama, and the conflating of quite distinct historical formations by the use of a particular twentieth century vocabulary) Cooke discusses the role of John Winthrop in the growth of Massachusetts. Speaking from the pulpit steps of a seventeenth century church, Cooke, with all his relaxed persuasiveness, goes on:

*That *Upstairs, Downstairs* struck a deep chord in many British television viewers did not go unnoticed by the advertising industry. The actor playing the canny Scot, Hudson (Gordon Jackson) went on to advertise a savings bank and Richard Bellamy's accumulated authority, as interpreted by David Langton, was deployed in the promotion of a dictionary—striking confirmation of Raymond Williams' insight of television as 'flow'.

. . . what started out as a trading company organised for profit turned into a religious dictatorship, and looking back on it we can see that it does bear a striking resemblance to a communist or other totalitarian government. The party will tolerate you as a citizen but you can have no say in the government if you are not a party member. Life is very real and very earnest and you're building a society from the ground up which cannot be sustained by consent but only by the most rigid discipline imposed from on top. So, the rule is, direct every function of society: work, play and worship and morals and business and literature. And in this case usually with admonitions from the Old Testament interspersed with the thoughts of Chairman Winthrop . . .

To set us aright, Cooke then proceeds to Philadelphia, to Benjamin Franklin and to what, for Cooke, represents the Holy Grail of history, that charter for the bourgeoisie — the Declaration of Independence.

10 Towards alternatives . . .

Ideological struggle is not like other forms of struggle. The only method to be used in this struggle is that of painstaking reason and not of crude coercion. (Mao-Tse-Tung)

The preceeding chapters have sought to demonstrate that the dominant practices of history-writing and television production (both individually and in mutually reinforcing ways) allow free passage to philosophical categories and aesthetic structures congenial to the maintenance of the system of social relations of advanced capitalism. In so doing, these dominant practices suppress, push to the margins, allow only limited currency to, alternative philosophical categories and aesthetic structures which are uncongenial to that system.

This chapter tries to indicate the extent to which alternatives to the dominant practices have been allowed currency in the juncture between television and history in Britain, what such alternatives look like and where further examples of alternative practices might be sought.

Inherent in the characteristic bourgeois separation between art and social life is the view that so long as the media-worker is clearly seen to be producing fiction, then what he/she does is of little political consequence. The moment, however, that the forms he/she uses cease to be unambiguously 'fictional' and begin to look like the 'factual' productions of the media, then he/she is seen to pose a political threat.

It should come as no surprise, therefore, that — as has been indicated earlier — radical historians have been rigorously excluded from participation in 'factual' programmes about history and that the broaching of alternative views of history has come from workers in the area of television drama seeking to extend and render politically relevant the constraining forms of bourgeois television drama. In the examples discussed below, therefore, the emphasis will be upon the extent to which it is necessary to transgress the dominant *aesthetic* forms of television in order to transgress the dominant bourgeois conception of history.

An early example of active reflection on the ways of rendering history on television is provided by *Culloden* (BBC), transmitted in 1965. One of the premises on which *Culloden* is constructed is the supposition: what if the resources and techniques of television had been available in 1746? However, it does not present itself as a simplistic 'window on the world' of 1746: it is a programme with a clear position on the events it describes, as the rubric following the title indicates:

> An account of one of the most mishandled and brutal battles ever fought in Britain.

An account of its tragic aftermath.
An account of the men responsible for it.
An account of the men, women and children who suffered because of it.

There follows a dramatic reconstruction of the battle and its aftermath with the techniques of documentary, *vox pop* and *vérité* television much in evidence, e.g., interviews with participants, a commentator on-camera (interestingly an eighteenth century figure — the biographer of Cumberland, the English commander — rather than a figure translated from another historical period) and an 'invisible' off-camera narrator whose narration (as well as carrying the pro-Highland, anti-English statement of the piece) provides 'hard' information about the precise times of the various phases of the battle, the precise weaponry and tactics used, and the precise clans and regiments involved and their casualties.

The features transgressive of bourgeois television drama operate alongside extensive retentions of many of its features: linear narrative; 'classical' composition and *mise en scène*; the careful orchestration of dramatic *crescendi*; and the central role accorded *individuals*. This mix proved both a critical and a popular success, *Culloden*'s originality lying in its deployment of well understood television procedures in a new context rather than in the creating of a new relationship between the events on the screen and the viewer.

The impulse of *Culloden* was an extremely generous and progressive one, the reminding of British audiences of events not far short of genocide which occurred barely two hundred years ago in this country. This, the rendering immediate of history, and the information it gives to a wide and diverse audience of other aspects of British history, are the strengths of the programme. The latter point is well illustrated by the early scene in which — in the most characteristic motif of this and others of Peter Watkins' works — the camera lingers on *faces*, in this case the faces of the clansmen, and the narration proceeds:

> . . . Angus Macdonald, servant of a sub-tenant. He owns nothing. Lowest in the clan structure, he is called a 'cotter'. This man is totally dependent on the men above him in the clan system. They, in their turn, on the tacks men. They, in their turn, on *this* one man: the man who has brought them all onto the moor. Alexander Macdonald . . . Chief of the Macdonalds of Keppoch, the owner of all his tenants' land. The rent he has charged them is to fight with him as clan warriors whenever he decrees. This is the system of the highland clan — human rent.

As a statement, in a dramatic programme aimed at a wide audience, of the social relations of feudalism, this is more than adequate. What, then, might be the limitations* of *Culloden* as an alternative to the dominant practices in

---

*By raising the issue of the possible limitations of the practice exemplified by *Culloden*, I am not for one moment impugning its progressive qualities. The same practice, deployed in Peter Watkins' next work, *The War Game* — an account of a nuclear attack on South East England and an examination of the feasibility of existing defence arrangements — caused sufficient discomfiture to government for it to be banned by the BBC. The strength of that practice however, is agitational rather than analytic.

television- and history-writing? These have to do with its central impulse —
deriving from the dominant concerns in television as set out by Stuart Hall in
chapter 3 — to give viewers the sense of actually being there at Culloden and its
aftermath; its consequent impulse to make viewers *feel* rather than *think*
history; and its encapsulation of itself strictly within the period and chronology
of the events it deals with. On this latter point, it has been remarked above
that in designating a participant as on-camera reporter of the battle, the
programme opted for the contemporary figure of Cumberland's biographer —
thus retaining the period as conceptually distinct from the twentieth century —
rather than injecting a modern figure with all the knowledge and historical
perspective of his own time. One has the feeling that the makers of the
programme would have regarded the latter course of action as an *unacceptable*
transgression of television's 'laws', even though such a transgression is present,
albeit invisibly, in the off-camera interviewer of Prince Charlie and the other
figures at Culloden.

The voice-over narration compounds this limitation. Immensely flexible (in
terms of handling concepts) as narration may be, the narration of *Culloden*
makes little attempt to locate the meaning of *Culloden* within the historico-
political forces of the modern world. To be sure, the narration locates the battle
in a context wider than itself:

> Thus has ended the last battle to be fought in Britain and the last armed
> attempt to overthrow its king. The establishment has been saved, peace
> restored, church, crown, trade and commerce safeguarded.

However, the key concepts of which the programme is virtually innocent are
mode of production, uneven development, colonialism and *imperialism.* Although,
as indicated above, the programme refers to the clan system and how it
operates, there is no sense that the clash at Culloden is meaningful within the
struggle between feudalism and nascent capitalism spread over several centuries
in Europe and that the 'vigorous police action' of Cumberland's troops after
Culloden, leading to the near-extirpation of Gaelic culture, is meaningful in
the context of Europe's relationship with the Third World over the same
centuries.

*The key point to be noted is that for the programme to have foregrounded
these historico-political concepts, it would have had substantially to recast its
aesthetic strategies.*

Perhaps the most controversial alternative project involving television and
history was the BBC's 1975 transmission of *Days of Hope,* a series of four
ninety-minute plays dealing with the decade 1916-1926, a period chosen by
the makers (who included writer Jim Allen, producer Tony Garnett and
director Ken Loach) on account of its centrality in working-class history.
Among the important constituents of this period were the aftermath of the
Russian Revolution, the First World War and what working-class attitudes to it
were, the Irish rebellion, and the General Strike of 1926. The debate which the
makers hoped would ensue from the transmission of the films was that about
reformism or *revolutionism* as the means to working-class power. In the event,

the quite sharp public debate which followed the transmission centred on issues of factual accuracy ranging from whether soldiers in 1916 wore their equipment this way or that to whether particular senior figures in the Labour administration and Trades Union Congress conspired with rightist forces against working people; and on the issue of whether such an obviously 'political' work should have been transmitted at all. In the latter context William Deedes, the editor of the right-wing *Daily Telegraph*, in a discussion on the BBC's *Tonight* immediately after the last programme in the series, advanced the argument referred to above: that *Days of Hope* was confusing as to whether it was 'art' or history and that such ambiguity ought not be to permitted. Needless to say, no such argument was advanced against, say, *Edward the Seventh* nor did any public debate follow its transmission. The lesson to be learned, of course, is that programmes which support the dominant ideology are regarded as natural and the few which do not are regarded as political.

The transmitting of *Days of Hope* (as of all the work of Allen, Loach and Garnett) was an extremely important event which has not received, particularly on the left, the analysis it deserves, an analysis which would have to include the specific aesthetic strategies of *Days of Hope*, the extent to which it is similar to the bourgeois television drama which surrounds it and the ways in which it is significantly different. Some sense of what this kind of analysis might look like is provided by Raymond Williams in his remarks on another Allen/Garnett/Loach play, *The Big Flame* (BBC), in his article 'A Lecture on Realism',[1] a quietly polemical piece which, as Williams puts it, seeks:

> to take the discussion of realism beyond what I think it has been in some danger of becoming — a description in terms of a negation of realism as single method, of realism as an evasion of the nature of drama, and the tendency towards a purely formalist analysis — to show how the methods and intentions are highly variable and have always to be taken to specific historical and social analysis . . .[2]

Clearly, one of the perspectives Williams has reservations about is that of the journal *Screen* whose position on realism is described in chapter 5. Williams' warnings notwithstanding, certain elements of realist practice and, within this, of *Days of Hope*, remain highly problematic. As Colin MacCabe writes:

> In order to fracture this unity (whereby knowledge of the truth is guaranteed) it would be necessary to pose the problem of the conditions of representation; it would be necessary to interrogate the reality of the constitutional tradition which allows films like *Days of Hope* to be shown on television. To pose these problems would also and immediately pose the problems of the lessons of what happened then for the situation today — the transparent immediacy of the film would be broken by analysis. Only thus could the position of the viewer be fractured and with no obvious assigned position, he or she would have to work on the material. It could be objected, at once, that such a film would have a much smaller audience than *Days of Hope* managed to attract. But this raises the question (which does not seem to have been posed by the makers) of

who the play is addressed to. In so far as this question is not posed then the film falls within a bourgeois conception of history in which the past is understood as having a fixed and immutable existence rather than being the site of a constant struggle in the present. And it is this conception of history which places *Days of Hope* firmly within the most typical of the BBC's varieties of artistic production: the costume drama. Another feature of this lack of analysis is the context of the knowledge that Ben (one of the main characters) and the viewer have gained by the end of the film. Given the fact that this knowledge is final, which is a necessity imposed by the form, and given that the General Strike was a failure, a necessity imposed by history, the only knowledge that the text can produce (which will have the necessary finality and leave history as it is) is that of betrayal. Given that we can see that the working-class were honest, straightforward and committed to socialism, their defeat must be the work of leaders who betray them. And this, of course, raises the question of the film-makers' political sympathies and affiliations.[3]

Among the substantial issues raised by MacCabe about *Days of Hope*, the concession appears to be made that the kind of text MacCabe is canvassing in his critique of the classic realist text, the kind of text which provides the necessary space for the operation of a conceptual apparatus, must of necessity attract a smaller audience than works such as *Days of Hope*. That conclusion is premature on the evidence of what is, in many respects, the most interesting attempt thus far to unite television and radical historiography in a dramatic mode which promotes both pleasure and analysis — the television adaptation of the 7:84 Theatre Company's play *The Cheviot, The Stag and the Black, Black Oil* (BBC).[4]

It is evident, even from the sequences surrounding the programme's opening titles, that it is transgressive of one of the dominant features of both bourgeois historiography and bourgeois drama — encapsulation within a single historical period. There is, surrounding the main titles, a montage of shots: a truck on a building site; a highlander being chased through the heather by redcoats; an oil explosion at sea; an oil rig; sheep grazing; a gentleman (wearing shooting gear of an indeterminate period of the last hundred years) shooting deer; and a worker shooting a verey pistol to ignite a gas jet on an oil rig. This montage poses a conceptual relationship, which the programme will develop, linking: the Highland Clearances of the first half of the nineteenth century whereby crofters were moved out to make room for the more economically productive Cheviot sheep; the development of Highland game parks in the second half of the nineteenth century; and the exploitation of the Highlands in the off-shore 'oil boom' of the 1970s.

The sequences surrounding the main titles signal something else which is crucially important: something which also complicates the temporality of the programme (cf. chapter 7) and at the same time signals that it is not simply a television adaptation of the play but an artefact which joins television adaptation with a *specific* performance before an audience of highlanders in Dornie in the western highlands. The interaction between the events happening

51

on the stage (and by imaginative extension the events presented in more traditionally televisual terms) and the specific audience is a key element in the force of the piece.

Another dominant feature both of bourgeois historiography and bourgeois drama which is transgressed is the autonomy and continuity of the individual consciousness. In *The Cheviot, the Stag and the Black, Black Oil* this takes the form of the abandonment of the practice whereby particular actors or actresses are encapsulated within a particular 'character' with the consequent focus of interest being the development of that character and the performance of the actor/actress, and its replacement with a practice which has more to do with the demotic forms of circus and music hall whereby actors and actresses assume a variety of roles within the space of the work according to the requirements of the sub-scene.[5] Thus, in the course of *The Cheviot, The Stag and the Black, Black Oil*, a particular player may fulfil the roles of narrator, singer, scene-shifter, nineteenth century land speculator, twentieth century property speculator, and Texas oil man. Crucially, the processes of giving direct pleasure to the spectators (largely through songs and sketches) and the requirements of political reflection take precedence over the display of actorial virtuosity which is such a central impulse in the bourgeois (particularly British) theatre, cinema and television.

An important feature of dominant television practice, voice-over narration from which the political stance of the piece is adumbrated and reinforced (cf. chapter 5), is retained but its content is rendered progressive and analytic. The discussion of *Culloden* above pointed to the lack of certain key concepts such as *mode of production, uneven development, colonialism* and *imperialism* in its handling of the events of Culloden and its aftermath. These concepts are strikingly present in *The Cheviot, the Stag and the Black, Black Oil*, in the songs, the humorous sketches, the historical reconstructions and, pre-eminently, in the narration which never loses sight of the relationship of the Clearances, the Game Parks and the 'oil boom' to each other and to historical forces and phenomena outside Britain itself. Thus, a narration, which shifts from player to player (underlining the demotic, ensemble-based practice of the company), goes on:

> . . . the Clearances gathered momentum. Hundreds of thousands of people were driven from their homes all over the north of Scotland. There is no doubt that a change had to come to the highlands. The population was growing too fast. The old methods of agriculture couldn't keep anyone fed. Even before the Clearances, emigration had been the only way out for some. But this coincided with something else. English, and Scottish, capital was growing powerful and needed to expand. Already huge profits were being made as a result of the Industrial Revolution and improved methods of agriculture and this accumulated wealth had to be used to make more profits, because this is the law of capitalism and it expanded all over the globe. And just as it saw in China, the Middle East, Africa, the West Indies, Canada, ways of increasing itself, so here, in the highlands of Scotland, it saw the same

opportunity. The technological innovation was there—the Cheviot, a breed of sheep that could withstand the highland winter and produce fine wool—and the money was there. Unfortunately, the people were there too . . .

and

Between 1810 and 1880 the landlords and speculators stocked the hills with sheep . . . somebody, somewhere had a safe return on investment. But the people had to go. They went to the appalling slums of Victorian Glasgow, on cholera-ridden boats to Canada, America, Australia, South Africa and they themselves drove out and subjugated other peoples for their land. The highland exploitation chain-reacted round the world . . . in Australia the aborigines were hunted like animals, in Tasmania not one was left alive; in America the plains were emptied of men and buffaloes and the seeds of America's imperial power were firmly planted . . .

There is a tendency among those seeking alternatives to the dominant bourgeois forms and practices to reject out of hand the whole catalogue of techniques and effects of bourgeois art and pose radical alternatives on a one-to-one basis. As an example of this, the central reliance of bourgeois art on dramatic climaxes and *crescendi* is felt to require, on the part of some radical practitioners, a commitment to severely cerebral structures and to forms of de-dramatisation.[6] This, of course, is a matter to be decided within the overall strategy of particular works, but an across-the-board rejection of dramatic pacing and climax should be viewed with great caution. The cerebral dimension of *The Cheviot, the Stag and the Black, Black Oil* has already been demonstrated, primarily by reference to its montage, actorial and narrational strategies. *Within that particular mix* its use of very traditional dramatic forms is telling, as in the sequence, a historical reconstruction on film, in which the Duke of Sutherland exhorts his tenants to enlist in his own regiment for service in the Crimean War. The sequence is structured round dramatic reversals; the confident address by the Duke; the sullen silence of his tenants; the Duke's desperate offer of six golden sovereigns to every man who enlists; the continued sullen silence; the Duke's vicious attack on their 'cowardice'; and the carefully controlled pacing within which an elderly and dignified tenant rises to address the Duke, culminating with the words:

It is the opinion of this country that if the Czar of Russia should occupy Dunrobin Castle we could not expect worse at his hands than we have experienced at the hands of your family in the past fifty years. We have no country to fight for. You robbed us of our country and gave it to the sheep. Therefore, since you prefer sheep to men, let the sheep defend you.

It is difficult to resist the conviction that, *in an appropriate mix of methods and techniques designed to foreground conceptual issues and provoke reflection,* traditional strategies executed with the force of the sequence referred to above must retain a place.

The discussion in this chapter has focussed on only *three* putative alternatives to the massive dominance of bourgeois categories in relation to history on British television. Needless to say, there are other examples worthy of analysis, e.g., the television work of Trevor Griffiths, particularly his contribution to on-going series such as *A Fall of Eagles* (BBC)—the object of analysis in this case being to see how a sophistricated Marxist dramatist negotiates predominantly bourgeois forms.

Equally, there are practices outside Britain which require understanding and interrogation. These include that of Straub/Huillet (already referred to); the extensive series of historical biographies made for television by Roberto Rossellini; and (what promises to be an interesting body of work dealing with the conjunction of cinema and history) the films of Jean-Louis Comolli. The practice of all these figures who have chosen to deal directly with questions of history and historiography constitutes, of course, a sustained reflection on the relationship between aesthetics and ideology, a process of reflection which is carried on by media workers the subject matter of whose work is other than history.[7] While recognising the importance of any challenges to the dominant way of constructing and depicting reality, this monograph has sought to clarify the conjunction television/history in the belief that such a clarification is particularly urgent in relation to the political struggle in Britain in the present conjuncture, a struggle which manifests itself in class, sexual *and* ethnic terms.

This monograph, with all its limitations, is an engagement on the ideological front of that struggle. It is not the front on which the war will ultimately be won, but it is an essential battleground nevertheless.

Notes

1. *Screen*, Vol. 18, No. 1, Spring 1977.
2. Ibid., p.73.
3. Colin MacCabe, 'Days of Hope—a Response to Colin McArthur' in *Screen*, Vol. 17, No. 1, Spring 1976, pp.100-101.
4. The theatrical practice represented by the work of the 7:84 Company, and its implications for radical cinema and television practice, ought really to be discussed in relation to the theory and practice of Bertolt Brecht—a praxis centrally concerned with the relationship of the spectator to the events represented and the creation of a space for analysis and reflection. Lack of space precludes such a discussion here, but its implications for radical cinema and television practice can be followed up in the special issue of *Screen* (Vol. 16, No. 4, Winter 1975/76) devoted to Brecht. One of the *exempla* of Brechtian practice cited therein is the Straub/Huillet film *History Lessons*, the complete script of which is printed in *Screen*, Vol. 17, No. 1, Spring 1976. *History Lessons* is highly relevant to the discussion in this chapter.
5. For the other possible dimensions of such an alternative practice see Peter Wollen, 'Counter-cinema: *Vent d'Est*' in *Afterimage*, No. 4, Autumn 1972.
6. See the work of figures as disparate as Jean-Luc Godard, Straub/Huillet, Yvonne Rainer and Chantal Akerman.
7. A useful account of this process of reflection is contained in Peter Wollen, 'The Two Avant-Gardes' in *Studio International* Nov-Dec 1975.

Appendix One: A Note on 'Popular Memory'

As has been pointed out elsewhere in this monograph, there has been a movement among radical historians in Britain and elsewhere to bring to light working-class struggles of earlier times with a view, of course, to providing models for struggle in the present day. This impulse in the British context was in reaction to the virtually exclusive concern, in the dominant historiography, with the aristocracy and the bourgeoisie and their institutions and so the movement in the British context tended to be known as 'history from below'.

A somewhat analogous project got under way in post-1968 France, but out of a very different context. Many intellectuals in France, not sharing the French Communist Party's (PCF) analysis of the May events and, in particular, its view that they did not constitute a revolutionary situation, were extremely dismayed by what they saw as the Party's failure of nerve at this time. From left positions outside the Party, these intellectuals mounted a critique of the PCF, a central strand of which was in relation to the PCF's version of history, including the history of the French working-class. It was in this context that the notion of Popular Memory was posed, a popular memory of struggle which was outside the institutions of the PCF.

This impulse found its way into the film journal *Cahiers du Cinéma* in 1974 in the context of an attack on 'la mode rétro', a style of film-making which used recent history, but displaced political questions onto the terrain of the personal, particularly the sexual (e.g., films such as *The Night Porter* and *Lacombe Lucien*) with the result that the actual struggles against fascism which constituted the subject matter of these films were repressed. This was seen as paving the way ideologically for a post-Gaullist alliance under Giscard d'Estaing between the nationalist right of de Gaulle and the collaborationist right of Vichy.

The most coherent articulation of the Popular Memory position has been in the work of Michel Foucault in works such as *Madness and Civilisation, The Order of Things, The Archaeology of Knowledge,* and *The Birth of the Clinic.* Foucault is interested in the analysis of forms of discourse. He understands very clearly that the past is irrecoverable and unreconstructable and that historians have to construct *knowledge* (of the past but *for* the present) on the basis of what the past said about itself in its forms of signification, most notably writing. The method is perhaps clearest in the collection of texts *I, Pierre Rivière* made by Foucault's seminar, and edited by Foucault himself, in which the memoir of Pierre Rivière (a nineteenth century peasant who murdered his mother, brother and sister) is placed side by side with medical and legal reports, newspaper accounts and eyewitness testimony, revealing a struggle of discourses which hinges on the relative power of institutionalised forms of knowledge in nineteenth century France.

Whatever their differences of context, the British History from Below movement and the French Popular Memory impulse have the effect of fore-grounding the experience of levels of society usually suppressed by dominant historiography. As such their effect is progressive and, in the suffocatingly patrician context of the United Kingdom, particularly necessary. However, by bringing together the ideas of working-class and 'memory' they both pose the highly dubious notion of a class* as a unitary and timeless phenomenon, as an *identity* rather than—as it must be for the purposes of political action—an *oppositional entity* located in a matrix of economic, political and ideological forces in constant struggle.

From this perspective, therefore, the notion of Popular Memory may have useful agitational application, but its capacity to produce political knowledge and a guide to action seems limited.

*For an examination of the problematic nature of the concept *class* and of its diverse empirical applications see Alan Hunt (ed.), *Class and Class Structure*, Lawrence and Wishart, 1977.

Appendix Two: A Note on Co-production and Publishing Tie-ups

I had hoped, in the early stages of the writing of this monograph, to produce a substantial amount of data and discussion about the two phenomena of co-production and publishing tie-up, both of which figure extensively in many areas of television but particularly—it is my impression—in the area covered by this monograph. That the issue is raised in the form of a brief note is due to the fact that the broadcasting companies are markedly less than enthusiastic about entering into discussion with outsiders about the nature of their commercial practices and the implications of these for the form of their programmes. This note is therefore primarily to point to a problem and suggest the work which needs to be done.

Of the two phenomena, the implications of the tie-up between publishers and television companies are probably less worrying than those relating to co-production. The basis of the tie-ups are simple—the extraction of the maximum commercial advantage mutually between television companies and book publishers. If an intellectual who is also a 'personality' is writing thirteen television programmes the scripts of which can be recast into a book, then the audience for one will pull that of the other and if a television company is producing a historical drama series about a British monarch, it makes sense (both for the television company and the publishers) to re-issue the biography on which the series is based and stress the television series in the new promotion of the book. What needs to be known, however, as well as the commercial details of such contracts, is the extent to which editorial control of the material is shared and on what terms. There is also a need to know the extent to which books like *Civilisation*, *The Ascent of Man* and *America* are recommended reading in educational institutions and how they are contextualised.

The co-production phenomenon has also, primarily, an economic basis. Many British television companies are involved in this and, in the area of tele-history, the BBC has a consistent relationship with the American company Time-Life Inc.

Co-production has a particular appeal both to British broadcasting companies and foreign enterprises at this historical moment. The foreign enterprises can inject what the British companies most want in the current economic climate— hard cash—and in return they receive a low unit cost product from the oldest, most experienced and, economically, most efficient television production set-up in the world.

Many of the facts which require to be known about co-production are the same as for publishing tie-ups: the financial terms; the extent of the division of editorial control; the implications for the choice of topics and for the format and content of programmes.

There is, however, a particularly worrying feature which is not endemic to the phenomenon of co-production but which is associated with it. This relates to the structure of American higher education and, specifically, to the statistic that of the twenty-seven million students following courses through institutions of higher education, only nine million are geographically fixed to a campus. Television is a key element in the courses of the non-campus-based students and there is some evidence that the tele-history co-productions discussed in this monograph are finding their way into the curricula of diverse courses in the USA, if not as core material then certainly as 'pacers' — a kind of icing on the unleavened bread of serious study. Given the virtually exclusive commitment of these programmes to the bourgeois conception of history (and, closely associated with this, that they operate primarily as mechanisms of the institution television rather than as mechanisms of historical knowing), this fact is particularly worrying. There is a need to know precisely how these programmes are functioning within both the U.S. educational system and that of the U.K.

Appendix Three: The Availability of Television Programmes

Many teachers, at various levels of education and in various subject areas, see particular television programmes when they are first transmitted, regard them as potentially useful material for their courses, but are frustrated by the material's non-accessibility outside of (possible) repeat transmissions.

There are several reasons for this general non-availability of television material. Inside the television institutions themselves a distinction is made between *educational* or *schools'* broadcasts and the other programmes which constitute the bulk of what is seen by viewers. The television institutions obviously feel some responsibility to permit the continuous availability of the former (hence the current position whereby educational institutions are permitted to record these broadcasts off air and use them for three years).

However, in general no such responsibility seems to be felt by the television institutions with regard to what they would describe as their *non-educational* broadcasts despite the fact that it is them as much as schools' broadcasts that teachers are interested in. That this is so relates partly to the ideology of transience and disposability within which the broadcasting institutions regard their own products but, equally, it stems from the immense contractual difficulties the institutions face in attempting to extend availability beyond single transmissions, e.g., difficulties with regard to the copyright of actuality footage and the repeat rights of the various workers (actors, musicians, etc.) in drama programmes.

That these difficulties are not insuperable is evidenced by the willingness of the various television institutions to make some of their general programmes available through certain film libraries (e.g. the BFI Distribution Library, Concord Film Library, and Film Forum) and most particularly, by the work of the BBC's marketing organisation BBC Enterprises, which, from frankly economic motives, has negotiated the difficulties referred to above and offers some of its general programme material for sale and hire.

Clearly, however, there is a need for a mechanism whereby those parts of the general programme material of the television institutions which teachers most require may be negotiated for. Discussions about the creation of such a mechanism are currently going on in the British Film Institute.

Certain of the programmes discussed or referred to in this monograph are already available. For example, the series *America* and *The Ascent of Man* may be hired from BBC Enterprises Film Hire, Woodston House, Oundle Road, Peterborough and the series *World at War* from Film Forum (dbw) Ltd, 56 Brewer Street, London W1. The series *Civilisation* is available from the British Film Institute Distribution Library, 9 Chapone Place, off Dean Street,

London W1, and negotiations are under way to make available some of the other programmes discussed herein. If these negotiations are successful, an announcement will be made in *BFI News*.